Biosurfactants: Greener Surface Active Agents for Sustainable Future

Deepansh Sharma

Biosurfactants: Greener Surface Active Agents for Sustainable Future

Microbial Surfactants

 Springer

Deepansh Sharma
Amity Institute of Microbial Technology
Amity University Rajasthan
Jaipur, Rajasthan, India

ISBN 978-981-16-2704-0   ISBN 978-981-16-2705-7   (eBook)
https://doi.org/10.1007/978-981-16-2705-7

This Springer imprint is published by the registered company Springer Nature Singapore Pte Ltd.
The registered company address is: 152 Beach Road, #21-01/04 Gateway East, Singapore 189721, Singapore

Preface

The domain of microbial metabolites has been undergoing rapid changes in the last decade. The consumer is nowadays more concerned about the origin and sustainability. Now, industries are adopting the molecules under the purview of greener sustainability due to the guidelines of state regulatory agencies or as a green initiative to address the concern of consumers. Green labelling of industrial products is now a status symbol for society. This book is directed towards budding researchers, in biosurfactants research and chemical engineering. The present book mainly focussed starting from the basics of biosurfactants properties, group of microbes involved, qualitative and quantitative screening along with high-throughput approaches. The chapters highlight the merits of the surfactants obtained from the microbes over the chemically synthesized surfactants. Life cycle assessment (LCA) is incorporated in the book to get the scientific idea of the superiority of greener surfactants over chemical surfactants.

The various applications of biosurfactants are covered extensively, and the chapters are managed to reflect such applications in food, agriculture, and the environment. The emphasis throughout, however, is on the biodegradability, environmental sustainability, eco-friendly nature behind these applications. We would like to thank university management and especially Professor Gajender Kumar Aseri, Amity Institute of Microbial Technology at Amity University, Rajasthan, for their support, help, and encouragement, and Ms. Mehak Manzoor who helps to compile the bibliography content of the whole portions of the script. The author is grateful to his family for their patience during the writing compilation and preparation of the manuscript. Thanks to Dr. Deepti Singh for her constant and great support.

Jaipur, India   Deepansh Sharma

Contents

About the Author

Deepansh Sharma is an Assistant Professor senior grade and a Coordinator of the Funds for Improvement of Science and Technology Infrastructure (DST-FIST) at Microbiology Department, Amity University, Rajasthan. His primary research interests are in functional foods, probiotics, and novel holistic food formulations. He has completed his Ph.D. in the field of Industrial and Food Microbiology from the Kurukshetra University and Karlsruhe Institute of Technology, Germany with a scholarship from DAAD, Germany. Following his doctorate, he pursued his post-doctoral research at the Division of Dairy Microbiology, National Dairy Research Institute, Karnal, India. He has 12 years of research and teaching experience in food and Microbiology. He has produced 35 scientific publications, authored, and edited 7 international books and more than 30 book chapters. Dr. Deepansh Sharma has filed various patents related to microbial metabolites and controlled fermentation processes. He has been selected for the prestigious DAAD fellowship to work in Germany as a visiting researcher, young scientist award, and research excellence award during his tenure at various academic and research organizations. He was also selected as a Project Investigator by the Department of Science and Technology, the Government of India, and various food industries. Dr. Deepansh Sharma is presently working on the possibilities of biosurfactants as an active ingredient in the food processing industry. He is an active contributing member of the American Society of Microbiology and the Association of Microbiologists of India.

Biosurfactants or Chemical Surfactants?

1

Abstract

Microbial surfactants exhibit various advantages over synthetic or chemical surfactants. BSs are regarded to be "green" substitutes to surfactants of chemical origin. Microbial surface-active agents, which exhibit significant surface properties, are considered as "greener surfactants." The majority of the surfactants now available in the market for industrial and environmental applications are chemically synthesized. In the last decade, consideration toward the BSs has been raised, which is primarily due to their extensive range of functional characteristics and the diverse production potential of the producing organism. There is an emerging need for the sighting of new non-pathogenic strains with maximal production potential and exertions to commercialize through efficient downstream processing are vital to meet the future forecasts of BSs market size. Biodegradability is recognized as a vital feature, exclusion of these amphiphiles from environmental sites, and, henceforth, degradability is regarded as an important property when estimating the environmental hazard linked with surfactant use. Regarding a solo approach for comparing BSs and chemical surfactants, would not be appropriate as no one tactic would consistently measure characteristics, usage, and disposal. Furthermore, the reply to this present situation hinges on what we mean by "efficiency." For any surface-active agent "efficiency" can be demarcated in different ways, i.e. critical micelle concentration, surface tension, and emulsification potential (the lower the concentration, the improved the surfactants), decrease in air-water surface tension, emulsification (how efficiently they stabilize two different mixtures with diverse surface tension). Even though BSs have been considered as a promising substitute for synthetic surfactants for years, but they still need to deliver on their efficacy on various fronts. It is hopeful to see that there is still abundant space is available for basic research of the molecular interactions of their surface potential, and it will be a huge step frontward to be able to fit this information into a better inclusive considerate of the interactions among microorganisms and their applications in biosurfactants applications as a greener sustainable agents.

Keywords

Biosurfactants · Detergency · Surface activity · Wettability · Emulsification · Foaming

1.1   Introduction

Microbial surfactants exhibit various advantages over synthetic or chemical surfactants. Biosurfactants (BSs) are obtained from microbes (such as bacteria, fungi, actinomycetes, algae, and yeast) and have various benefits over the synthetic surfactants, like lower toxicity, biodegradable, significant ecological compatibility, effective foaming, highly specific and selective activity in extreme environmental systems such as temperature, pH, and extreme salinity. So BSs are regarded to be "green" substitutes to surfactants of chemical origin. Microbial surface-active agents, which exhibit significant surface properties, are considered as "greener surfactants" [1–5]. The majority of the surfactants now available in the market for industrial and environmental applications are chemically synthesized. In the last decade, consideration toward the BSs has been raised, which is primarily due to their extensive range of functional characteristics and the diverse production potential of the producing organism.

BSs are exceptional amphiphilic molecules (hydrophobic moiety long-chain hydroxyl fatty acids) and hydrophilic moiety (peptides, carbohydrate, amino acids, phosphate, carboxylic acid, alcohol) with characteristics which have been used for a range of industrial and environmental applications [6, 7], pharmaceutical, food formulation, microorganisms enhanced oil recovery (MEOR) [1, 8–10], and bio-medical applications [11, 12]. BSs are categorized by distant structural multiplicity and have the potential to decrease surface tension and interfacial tension. The global surfactant market size is expected to reach USD 52.4 billion by the year 2025 from a total worth of USD 42.1 billion in 2020, at a CAGR of 4.5% (Global Forecast to 2025). Edible bio-emulsifiers or surfactants would spread over to various applications together with food, cosmetics, environmental clean-up, healthcare, and industrial purposes. Rhamnolipid could be a latent substitute for the chemical-based surfactants and a vital molecule with a market worth of USD 2.8 billion in 2022. There is an emerging need for the sighting of new non-pathogenic strains with maximal production potential and exertions to commercialize through efficient downstream processing are vital to meet the future forecasts of BSs market size.

Moreover, increasing awareness about the new healthcare and sanitizing products such as skin sanitizer without ethanol due to COVID-19 is an added factor driving its demand. Besides, most of the generally adopted chemical surfactants are harmful to the skin in long-term exposure [1, 13] and readily biodegradable, their utilization may result in the buildup of recalcitrant ecologically detrimental substances in soil [14].

In the literature, data reveals that more than 350 patent were obtained and tally increases on BSs so far [15]. The microbial BSs have been advantageous over plant-derived surfactants due to the ease and time of the scale-up, rapid production, and multifunctional potential.

1.1.1 Classification of BSs Agents

Typically, BSs are known for the following characteristics advantageous for industrial use:

- Excellent surface-active agents
- Environmental compatibility
- Antimicrobial in nature
- Biodegradability in environment
- Production from renewable substrates

BSs are typically classified based on origin, producing organisms, and molecular weight like glycolipids, phospholipids, fatty acids, lipopeptides, and polymeric BSs [6, 8, 16, 17]. Synthetic surfactants were categorized based on the constituents and the polarity of the hydrophilic head group such as non-ionic, anionic, cationic, and amphoteric. But microbial BSs are categorized on a different approach, one option is to categorize the molecules by size, such as low-molecular-weight (LMW) BSs as glycolipids contrasted with high-molecular-weight (HMW) BSs as lipopolysaccharides (Fig. 1.1) (Table 1.1). In a different approach, LMW BSs is by net charge, therefore consenting for the diversity of non-ionic, anionic, cationic, and amphoteric BSs (Fig. 1.2).

All these different approaches to categorize BSs have been established due to their vast structural multiplicity, which would not fit a single classification method. It is quite interesting to know that, one type of BSs hardly ever comprises only one single molecule as a structural monomer. Instead, most BSs are obtained as different congeners, which have variations in the type of side chains, length, branching, and degree of hydroxylation. Various congeners were reported for well-recognized BSs, e.g. rhamnolipids and surfactins. On the other hand, HMW BSs have different complexities of the molecules as a multicomponent mixture of hetero-polysaccharides, lipopolysaccharides, lipoproteins, or combinations thereof. Similar to LMW BSs, HMW molecules can form emulsions of two immiscible suspensions. Although LMW surface-active agents are less effective in reducing surface tension, LMW is highly effective as bio-emulsifiers to stabilize cosmetic preparations, bakery products, and pharmaceutical products.

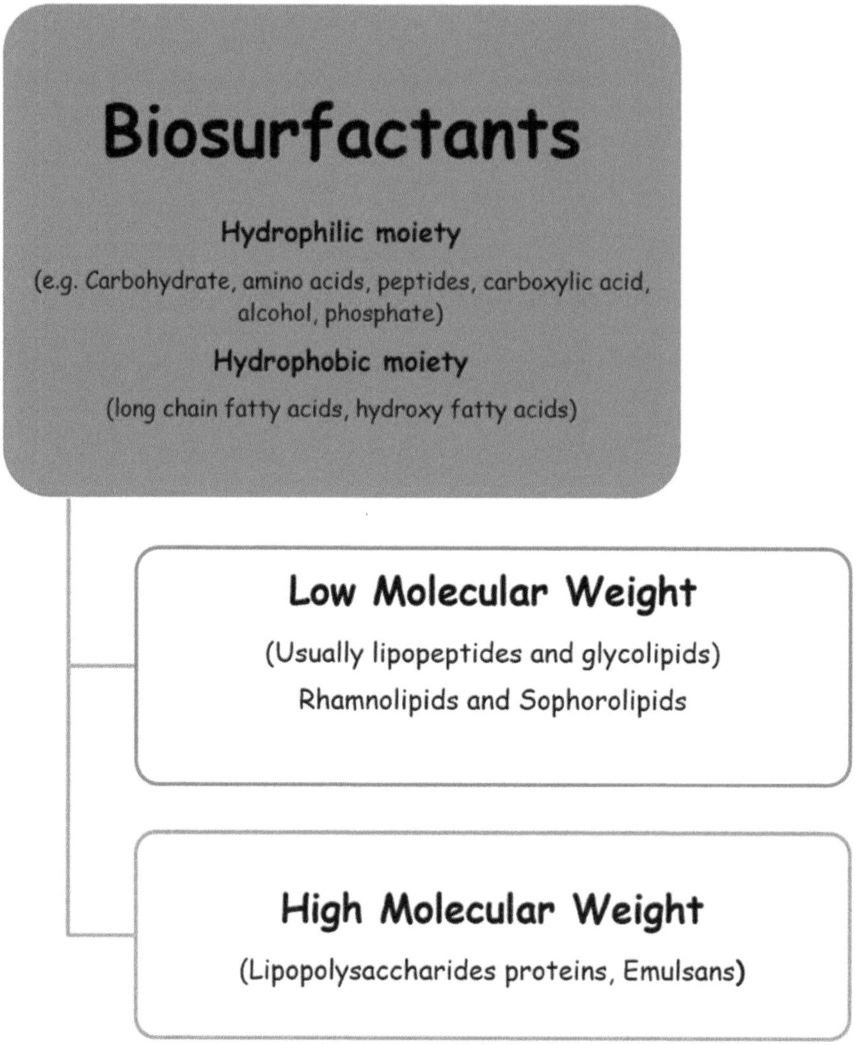

Biosurfactants

Hydrophilic moiety

(e.g. Carbohydrate, amino acids, peptides, carboxylic acid, alcohol, phosphate)

Hydrophobic moiety

(long chain fatty acids, hydroxy fatty acids)

Low Molecular Weight

(Usually lipopeptides and glycolipids)

Rhamnolipids and Sophorolipids

High Molecular Weight

(Lipopolysaccharides proteins, Emulsans)

Fig. 1.1 Classification of the biosurfactants (Courtesy: [1])

1.2   Low-Molecular Weight (LMW) Biosurfactants

1.2.1   Glycolipids

The important feature of a glycolipid is the presence of a monosaccharide or oligosaccharide attached to a lipid molecule. More than 250 diverse glycolipids composed of sugar and lipid molecules are the best-studied BSs as the LMW surface-active agents [19]. Diverse sugar moieties like glucose, aldohexose,

Table 1.1 Biosurfactants major group and producing microorganisms

Type of biosurfactants	Microorganisms
Glycolipids	
Sophorolipids	*Candida bombicola, Torulopsis petrophilum, Torulopsis apicola*
Mannosylerythritol lipids	*Pseudozyma hubeiensis, Pseudozyma tsukubaensis,*
Rhamnolipids	*Pseudomonas aeruginosa, Pseudomonas chlororaphis, Burkholderia glumae, Burkholderia thailandensis*
Trehalolipids	*Mycobacterium tuberculosis, Rhodococcus erythropolis, Rhodococcus actinobacteria, Arthrobacter sp., Nocardia sp.*
Cellobiolipids	Ustilago sp
Lipopeptides	
Surfactin	*Bacillus subtilis*
Serrawettin	*Serratia marcescens*
Subtilisin	*Bacillus subtilis*
Viscosin	*Pseudomonas fluorescens*
Iturin	*Bacillus subtilis*
Fengycin	*Bacillus subtilis*
Glycolipopeptide	*Lactobacillus acidophilus, Lactobacillus pentosus*
Particulate	
Vesicles and fimbriae	*Acinetobacter calcoaceticus*
Phospholipids	
Lipid phosphate	*Torulopsis magnoliae, Thiobacillus thiooxidans*
Polymeric biosurfactants	
Emulsan	*Acinetobacter calcoaceticus*
Liposan	*Candida lipolytica*
Mannan-lipid-protein	*Candida tropicalis*
Carbohydrate-lipid-protein	*Pseudomonas fluorescens*
Surface-active antibiotics	
Daptomycin	*Streptomyces roseosporus*
Entolysin	Pseudomonas sp.
Echinocandin	*Papularia sphaerosperma*

galactose, xylose, glucuronic acid, and rhamnose are widely known for their incidence. The lipid comprises saturated or unsaturated fatty acids, fatty alcohols, or hydroxy derived from fatty acids. Between glycolipids, to date, four different groups of BSs of commercial and environmental interest viz. rhamnolipids, xylolipids, sophorolipids, mannosylerythritol, and trehalose lipids are known popularly. Glycolipids are potential BSs because of their biodegradability, emulsification potential, and environmental compatibility.

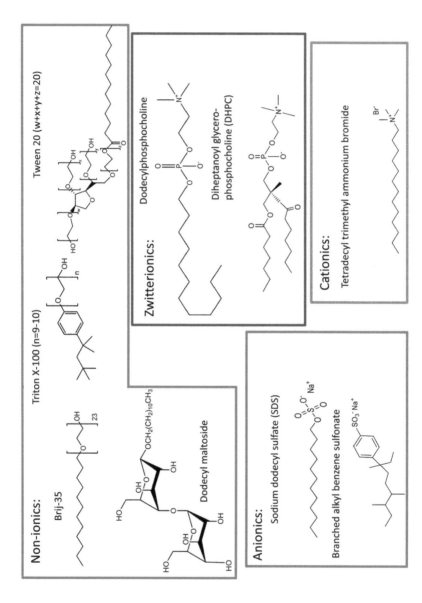

Fig. 1.2 Classification of BSs based on net charge (Courtesy by: [18])

1.2.2   Rhamnolipids

Various Pseudomonas species are reported for best known BSs molecule composed of rhamnose sugar known as rhamnolipids. Rhamnolipids are exhaustively reported microbial surface-active agents and belong to the class of glycolipids. Rhamnolipids have a hydrophilic glycosyl head moiety mainly with a rhamnose sugar and a 3-(hydroxyalkanoyloxy)alkanoic acid as a tail. Rhamnolipids contain 1–2 rhamnose sugar units with hydroxy fatty acids and are major to be obtained from *Pseudomonas aeruginosa* [20]. It was Bergström et al. [21] who has reported first-time rhamnolipids as "oil glycolipids." Rhamnose sugar is a major constituent of cell wall lipopolysaccharide in *P. aeruginosa* [22]. The biosynthesis of the rhamnolipids is catalyzed by the three different enzymatic steps [23, 24] (Figs. 1.3, 1.4, and 1.5).

1. RhlA is catalyzing synthesizing of HAAs [3-(3-Hydroxyalkanoyloxy) alkanoates], which is a hydrophobic precursor of rhamnolipids [27]

2. RhlB, i.e. rhamnosyl transferase utilized L-rhamnose and HAA precursors and synthesize mono-rhamnolipid [28]

3. RhlC, i.e. rhamnosyl transferase utilized mono-rhamnolipid as a precursor to synthesize di-rhamnolipid.

Operon containing genes that are responsible for the RhlA and RhlB production which is governed by regulatory genes, viz. RhlR and RhlI. The expression of RhlA and RhlB is controlling cell-to-cell signaling, i.e. quorum sensing [29]. It is a well-known fact that secondary metabolites and various antibiosis metabolites like antibiotics and proteases are typically regulated by quorum sensing [24].

In addition to Pseudomonas several other species have been documented as an exclusive source of RLs such as *Acinetobacter calcoaceticus* [30], *Pseudoxanthomonas* sp. [31], *Enterobacter* sp., *Pantoea* sp. Isolating novel rhamnolipids producing species is advantageous from a biotechnological utility to replace opportunistic pathogens like *P. aeruginosa* to ensure safety and containment level.

1.2.3   Sophorolipids

It was Gorin et al. [32], who describe the sophorolipids extracellularly synthesized by *Torulopsis magnoliae* yeast, which was taxonomically corrected later as *Torulopsis apicola* or *Candida apicola*. Sophorolipids are composed of acetylated 2-O-β-D-glucopyranosyl-D-glucopyranose monomer bounded with β-17-L-hydroxyoctadecanoic acid [33]. Later on, sophorolipids of *Rhodotorula bogoriensis* were found different from the biosurfactant of *Candida apicola*, which varies in its hydroxy fatty acid fraction. Sophorolipids are undoubtedly one of the most effective BS's obtained from non-pathogenic yeast; comparable to the rhamnolipids, which is obtained from the opportunistic pathogen *P. aeruginosa*. Moreover, the yield and substrate conversion of sophorolipids is better than rhamnolipids can be attained. The hydrophilic moiety of the sophorolipid is composed of the sophorose and the

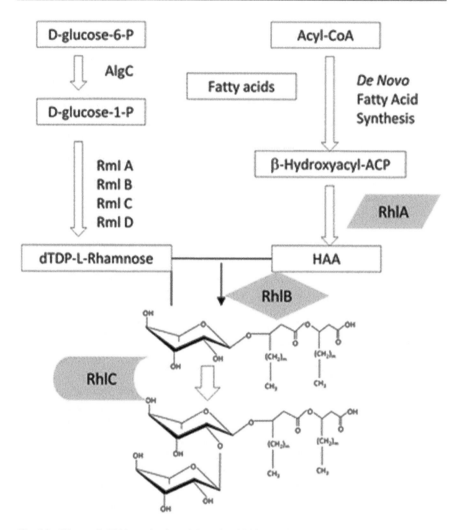

Fig. 1.3 Rhamnolipid biosynthesis and the role of RhlA, RhlB, and RhlC enzymes (Courtesy by: Qingxin [25])

hydrophobic moiety of the sophorolipid is made up of hydroxylated fatty acid on the terminal and subterminal position (Figs. 1.6 and 1.7).

The two different variations in sophorolipids majorly lead to the difference in physicochemical characteristics. Lactonized sophorolipids have diverse biological and functional characteristics as compared to acidic variation. Lactonic sophorolipids have an effective potential of reduction in surface tension antibiosis potential, however, the acidic variation shows an effective foam formation and solubility.

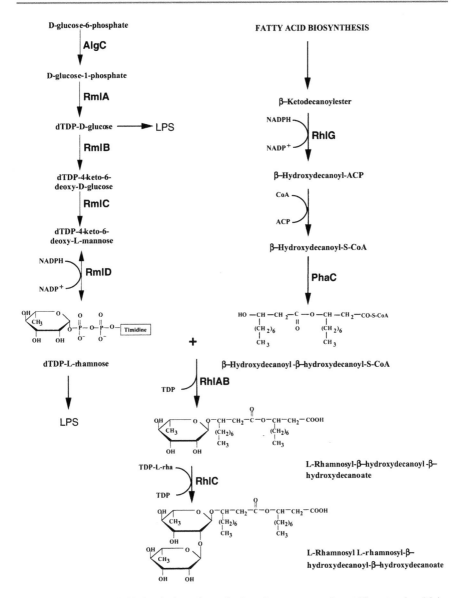

Fig. 1.4 Rhamnolipids biosynthetic pathway in *Pseudomonas aeruginosa* (Courtesy by; Maier and Soberon-Chavez [26])

Literature suggests that sophorolipids reduce surface tension in water from 72.80 to 30 mN/m with a critical micelle concentration ranging from 10 to 280 mg/L. More than a dozen species of yeast have been reported for the sophorolipid production such as *C. apicola*, *T. gropengiesseri*, *T. bombicola*, and *C. bogoriensis*.

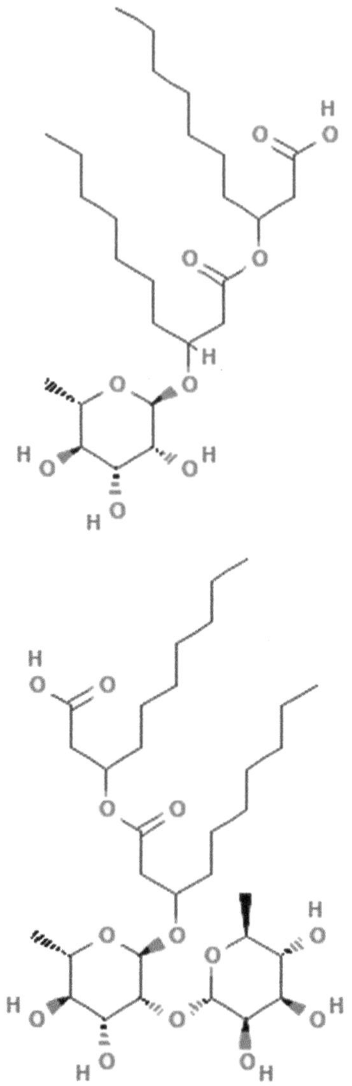

Fig. 1.5 Structure of mono
and di-rhamnolipid (pubchem
CID 5458394; 5,458,394)

1.2.4   Trehalose Lipids

The first trehalose lipid was reported in 1933 while studying *Mycobacterium tuberculosis* [35]. Presently different species of Mycobacterium, Rhodococcus, Arthrobacter, Nocardia, and Gordonia are known to produce trehalose lipids. Diverse structures have been determined predominantly in the *Rhodococcus* genus (Fig. 1.8). Trehalose is the elementary constituents of the cell wall of Mycobacteria and Corynebacteria species. Diverse trehalose comprising glycolipids are known to be obtained from various other microorganisms from mycolates groups, like

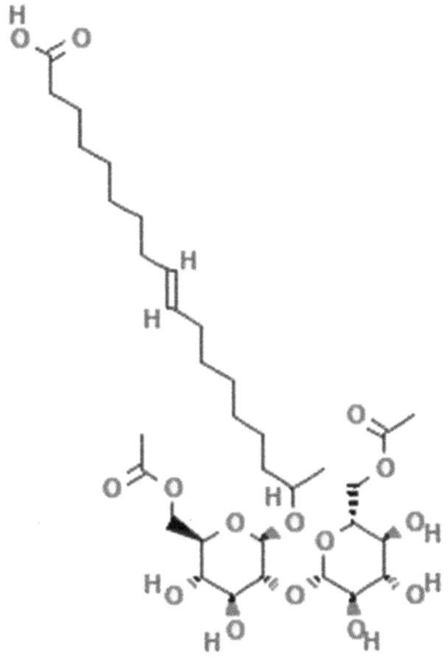

Fig. 1.6 Sophorolipids structure (PUBCHEM CID: 11856871)

Arthrobacter, Nocardia, Rhodococcus, and Gordonia. Such glycolipids vary in the chain length and vary from C20 to C90 of fatty acids esters. The majority of trehalose lipids produced by Rhodococcus and associated species are cell membrane associated.

1.2.5 Mannosylerythritol Lipids

Mannosylerythritol lipids (MELs) are a class of amphiphilic molecules with surface-active characteristics, categorized as glycolipid biosurfactants. Structurally, all four 4 available homologs have the same basic structure, which differs and are categorized as per their degree of acetylation: MEL-A is diacetylated at the position C4 and C6 positions, MEL-B and MEL-C are monoacetylated at C6 or C4, separately, while MEL-D has no attached acetyl group (Fig. 1.9) [37]. The above-mentioned structural variety confers to MEL excellent biological characteristics, which vary pointedly from the precursors [38]. MELs are a type of glycolipid BSs produced by Ustilago and Pseudozyma MELs contains mannose and erythritol as hydrophilic components and fatty acids and acetyl functional groups as the hydrophobic tail (Fig. 1.10). They can also be classified by the number of fatty acid chains and the chirality of MELs [39].

Different MEL-B producing species have been isolated from *Pseudozyma tsukubaensis*, using plant oil and glucose as major sugar. The different strains producing MEL-C namely *P. shanxiensis*, *P. hubeiensis*, *P. siamensis*, and *P. graminicola* are commonly reported. Additional research on basidiomycetes will likely be encouraged to search for other MELs with diverse structural modifications.

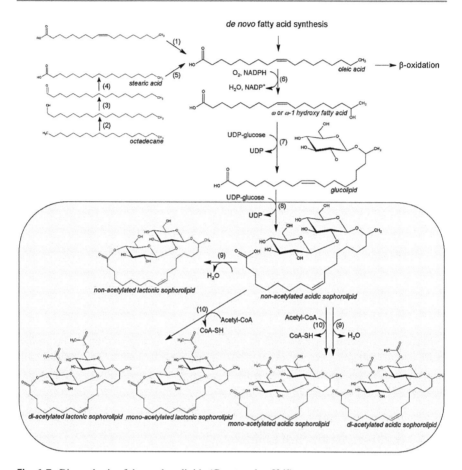

Fig. 1.7 Biosynthesis of the sophorolipids (Courtesy by; [34])

1.2.6 Lipopeptides

Lipopeptides are characterized as high structural diversity amphiphile with significant ability to reduce the surface and interfacial tension. *Bacillus subtilis* is known
for the production of different types of lipopeptides such as Surfactin, iturin, and
fengycin based upon amino acid sequence. Structurally, lipopeptides are composed
of a combination of fatty acid and a peptide fraction and link to an isoform group that
varies by the fractions of the peptide chain, fatty acid chain length, and the bonding
between peptide and fatty acids. These antibioses are either cyclopeptides in nature
such as iturins or macrolactones like fengycins and surfactins identified by the
occurrence of L and D amino acids and fatty acid tails (Fig. 1.9).

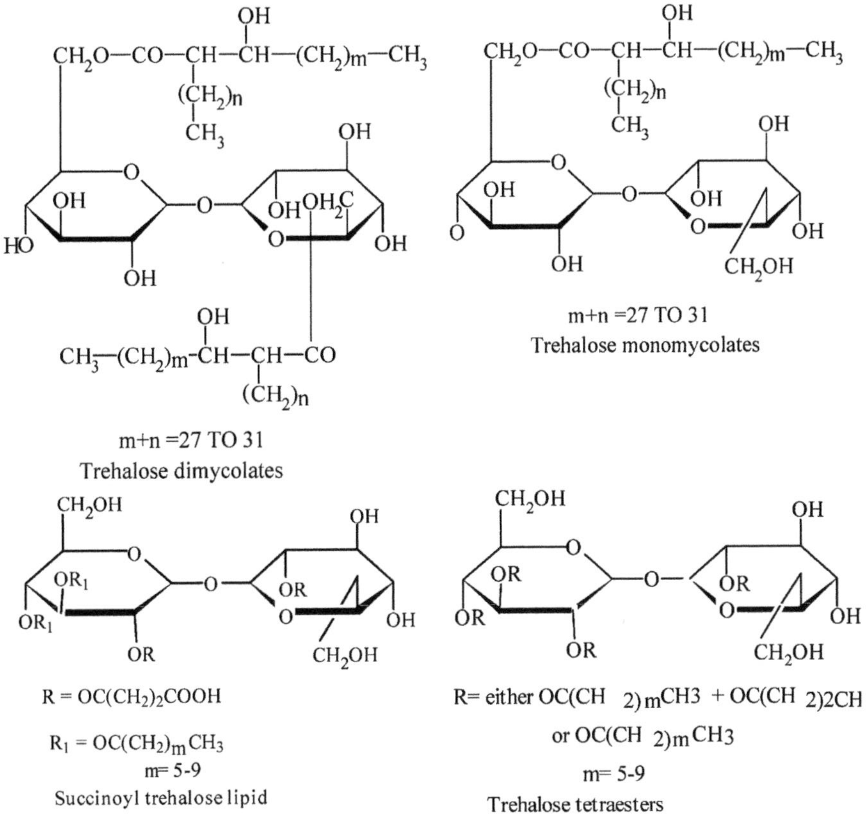

Fig. 1.8 Trehalose lipids along with side chains (Courtesy by: [36])

1.2.6.1 Surfactin

Surfactin is known as a bacterial lipopeptide with exceptional surface-active properties produced by *Bacillus subtilis*. Surfactin is a lipopeptide composed of 7 residues of D and L amino acids connected with a fatty acid chain of β- hydroxy fatty acid. The change in the amino acid composition of the peptide fraction is responsible to differentiate surfactin, lichenysin, esperin, and pumilacidin in the surfactin club [40]. Surfactin can reduce the surface tension of water from 72 to 27 mN/m. Surfactin is also known for its prominent applications as antibacterial activity and biological functionality (Fig. 1.11).

1.2.6.2 Fengycin

Fengycin is an antifungal cyclic lipopeptide amphiphile obtained from *Bacillus subtilis*. Fengycin inhibits the growth of filamentous fungi. Fengycins were reported to damage fungal hyphae; hyphae showed unconsolidated cytoplasm and damaged cell walls that were parted from the cell membrane [41]. Fengycin is structurally a type of lipopeptides containing 10 amino acids and a fatty acid connected to the

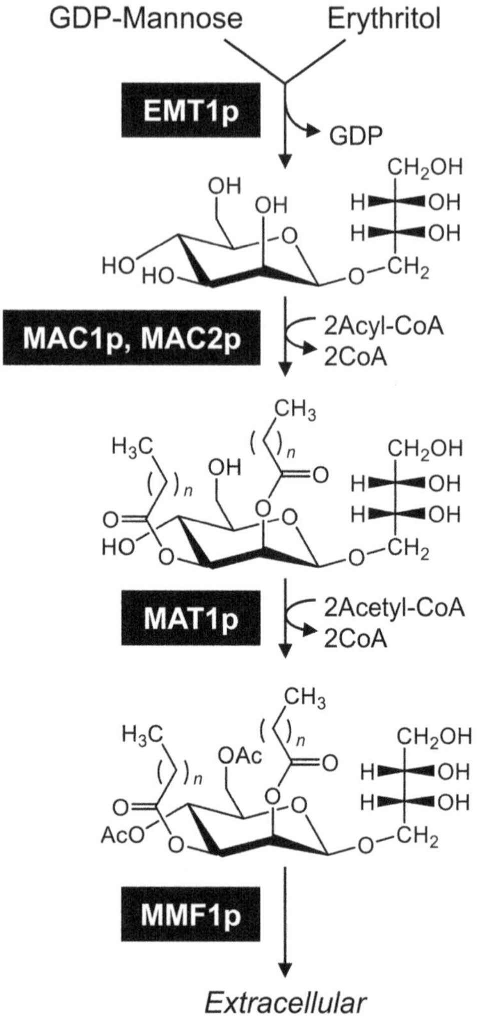

Fig. 1.9 Biosynthetic
pathway of MELs synthesis

GDP-Mannose   Erythritol

EMT1p

GDP

MAC1p, MAC2p   2Acyl-CoA
2CoA

MAT1p   2Acetyl-CoA
2CoA

MMF1p

Extracellular

N-terminal end. Strucurally fengycin containing unusual ornithine and allo-threonine as compared to the surfactin and iturin [42]. Fengycin composed is of 2 major moieties which differs in the composition of one amino acid. Fengycin A is containing 1 D-Ala, 1 L-Ile, 1 L-Pro, 1 D-allo-Thr, 3 L-Glx, 1 D-Tyr, 1 L-Tyr, 1 D-Orn, while in fengycin B the D-Ala is swapped by D-Val. The lipid fraction of fengycin A and B are more variable, as fatty acids have been recognized as anteiso-pentadecanoic acid, iso-hexadecanoic acid, n-hexadecanoic acid, and it can be a saturated and unsaturated fatty acid residue up to C18 [43].

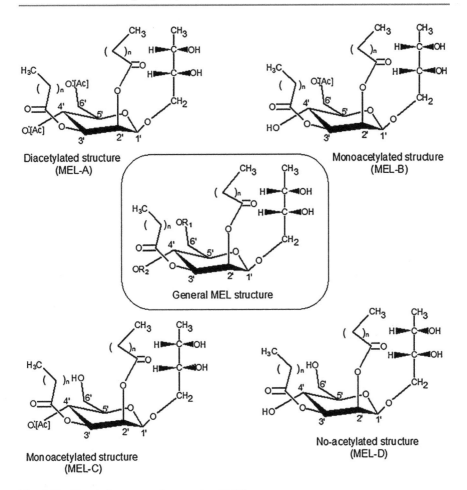

Diacetylated structure (MEL-A)

Monoacetylated structure (MEL-B)

General MEL structure

Monoacetylated structure (MEL-C)

No-acetylated structure (MEL-D)

Fig. 1.10 Chemical structure of conventional MEL

1.2.6.3 Iturin

Iturin A fits into a class of cyclic lipopeptides obtained from the cell-free supernatant of different species of *Bacillus subtilis* and which display antibiotic potential. Iturin A composed of a C_{14} or C_{15} amino acid concluding the LDDLLDL heptapeptide with a hydrophobic tail which contains 10–17 carbon atoms [44]. The iturin family, characterized by iturin A, mycosubtilin, and bacillomycin, are heptapeptides with a β-amino fatty acid which showed effective antifungal potential [45]. Iturin is insoluble in water and their biological property seems to be strictly related to their exchanges with cytoplasmic membranes [46, 47].

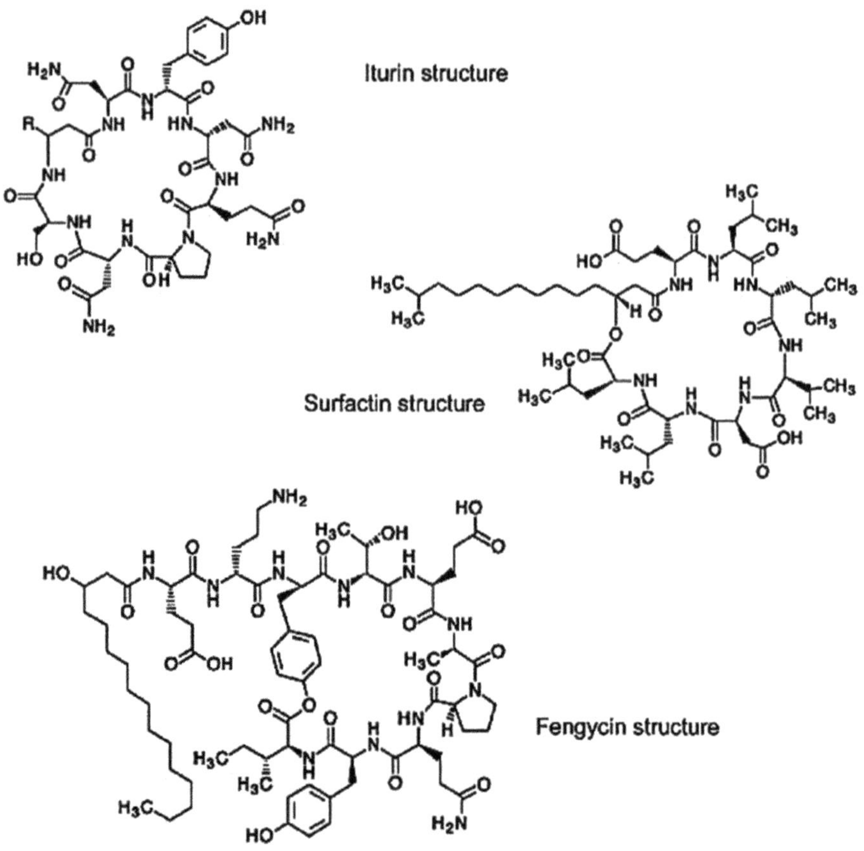

Fig. 1.11 Iturin, surfactin, fengycin structure

1.3   High-Molecular Weight BSs

HMW, polymeric BSs are recognized to be obtained from a diverse range of microorganisms. The HMW amphiphilic is the biological macromolecules that exhibit a high degree of structural complexity and can be mostly classified as (lipo)polysaccharides, (lipo)proteins, or compounds with combinations of these separate structural constituents.

1.4   Microbial Polysaccharide BSs

Emulsan is a well-characterized HMW polymeric surface-active agents. Polysaccharide amphiphile is known to be produced by various prokaryotes, while emulsans RAG-1 and BD4 are obtained from Acinetobacter (Fig. 1.12). Emulsan is an

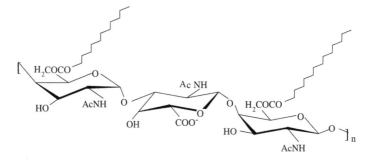

Fig. 1.12 Scheme of emulsan structure

extracellular polymeric bio-emulsifier produced by *Acinetobacter calcoaceticus* RAG-1. The unique structures of this class of polymers are their amenability to structural tailoring and their emulsification potential as low concentration as 0.001%). Emulsan is containing an unbranched sugar backbone with *O*-acyl and *N*-acyl connected fatty acid chains. The sugar backbone is composed of 3 amino sugars as D-galactosamine, D-galactosamino uronic acid, and a dideoxydiaminohexose in the proportion 1:1:1 [48].

Emulsion derived from Acinetobacter sp. composed of a multifaceted amalgamation of sugar and lipopolysaccharides, which comprises a 200–250 kDa polysaccharide with a positive charge. The lipid fraction composed of unsaturated fatty acid (C10-C18), and the amino moiety are characteristically acylated or bound to 4-hydroxybutyric acid mediated by amide bonds. Biosurfactant of *A. calcoaceticus,* i.e. BD4 exhibits a purposely diverse composition as related with RAG-1, it contains protein-polysaccharide contains heptasaccharides in a ratio of 4:1:1:1 of the monomers l-rhamnose, d-glucose, d-glucuronic acid, and d-mannose [49].

Another compositionally similar amphiphile which contains polysaccharides and fractions of proteins obtained from *A. radioresistens* and was recognized as alasane [50]. The emulsifying potential of the amphiphile is attributed due to the presence of 45 kDa alasane protein [51]. An additional HMF polysaccharide amphiphile was reported to be produced by *Pseudoalteromonas* sp. TG12 with a molecular weight higher than 2000 kDa containing xylose in the high ratio [52].

1.4.1 Microbial Protein Surfactants

Protein-rich surface-active agents are a class meagerly considered and currently only exhibit limited worth for industrial applications. Various other peptide-rich surface-active fractions are reported in different fungi [53]. Protein-rich biosurfactants are involved in the sequestration of metal contaminants [54]. *Emericella nidulans* is known for the production of protein-rich hydrophobins which have applications to stabilize cosmetics emulsions and other food emulsions. Hydrophobins were reported in 1991 as an amphiphile to produce hydrophilic and hydrophobic coatings at interfaces. Such biosurfactants are small cysteine-rich molecules containing 100 amino acid chain lengths. *Trichoderma reesei* was reported for the production of hydrophobins [55]. Hydrophobins are commonly present on the conidia surface and are recognized for their potential to form a steady hydrophobic layer.

1.5 Properties

1.5.1 Temperature, pH, Salinity, and Ionic Strength

Stability to environmental factors and industrial processing increased the suitability/ acceptance of BSs applications. Generally, BSs are considered for their significant stability over extreme pH, ionic strength of salts, and temperature abuse. The thermal stability, pH, and salinity of the BSs increase its scope of application in a larger perspective including the circumstances where elevated temperatures, acidic and alkaline pH, and salt concentrations prevail in most of the industrial and environmental applications (Table 1.2). *Oleomonas sagaranensis* and *Candida sphaerica* derived BSs exhibit a significant thermal and pH stability with reverence to surface tension decrease and emulsification potential. The BSs were found promising activity in terms of surface tension and emulsification at a high salt concentration [63]. It also displayed emulsification potential at a high ionic strength of salt concentration. The surface tension decrease and emulsification potential were comparatively stable at the high temperatures tested, demonstrating the appropriateness

Table 1.2 Biosurfactants properties and different application spectra

Properties	Biosurfactants	Application spectrum	Reference
Thermostable	Lipopeptides	Food and laundry formulations, oil recovery, antibiofilm potential	Ram et al. [56], Kiran et al. [57]
pH stability –Acidic range –Alkaline range	Glycolipid Lipopeptide	Food and biomedical Detergent formulation	Sharma et al. [58], Hmidet et al. [59]
Salinity	Glycolipid	Enhanced oil recovery (EOR), PAH degradation, detergent formulations for hypersaline environments, cleaning agents for food processing areas, remediation of wastewater containing salts, bioreduction of chromium, and other heavy metals	Derguin-Mecheri et al. [60]
Ionic stability to salts	Glycolipid	Nano-emulsions for food, flavor oils, cosmetics, and healthcare products	Bai and McClements [61]
Cold stability	Mannosylerythritol lipids (MEL)	Ice water interface	Kitamoto et al. [62]

of the BSs in the food, pharmaceutical, and cosmetics formulations, where the elevated temperature may be used to attain sterility of the product ([64, 65]). Additionally, BSs activity was affected by a change in pH and salinity from 2 to 14 and salt concentration varied from 0 to 15% (w/v). The surface tension of the BSs increased from 25 to 60 mN/m while moving from basic to acidic pH, but the emulsification potential showed a reduction from 69 to 55%. Additionally, there was no significant change in the surface activity with an increase in the salt concentration up to 16%. BSs obtained from Rhodococcus sp. showed stable emulsions over a wide range of pH 2–10, temperatures range 20–100 °C with an ionic strength of 5–25% salt (w/v). It makes it appropriate to utilize different perspectives in a diverse range of industries [6, 36, 66]. The stability of the BSs and robust behavior are very advantageous for applications such as microbial enhanced oil recovery and in situ removal of oil spills from the marine environment.

Temperature stable BSs have extensive significance in food, cleaning, and detergent formulations [1]. Novel thermostable BSs were obtained from Bacillus sp. SGD-AC-13 with a basal medium containing yeast extract with a short incubation period displayed excellent surface-active behavior, i.e. 31 mN/m at 160 °C, pH range of 6–12 [56].

The crude BSs obtained from Cryptococcus sp. demonstrated a significant reduction in surface tension and emulsification potential at 5–100 °C, pH (2–12), and elevated ionic strength (1–10%). The biosurfactant was displayed as a significant role in microbial enhanced oil recovery (MEOR).

1.5.2 Toxicity and Biodegradability

The majority of the microbial surface-active agents are obtained from bacteria, actinomycetes, yeasts, and molds during cultivation on different carbon sources [1]. BSs have indistinct advantages over synthetic surfactants, which include lower toxicity, a high degree of biodegradability, and improved ecological compatibility. Biodegradability is recognized as a vital feature, exclusion of these amphiphiles from environmental sites, and, henceforth, degradability is regarded as an important property when estimating the environmental hazard linked with surfactant use [67]. Biosurfactants are generally composed of sugar, lipid, amino acids, and proteins with phosphates. All these biological macromolecules are degradable by indigenous microflora of any ecological niche.

The environmental breakdown of the BSs molecules occurs involves two steps when microorganisms are involved and utilize it as carbon and energy sources. Initially, hydrocarbon breakdown occurs, leading to structural alteration and instantaneous forfeiture of amphiphilicity. Consequently, the intermediate product resulting from the first step is further transformed into CO_2, water, and elemental minerals [68].

$$BSs + degradative\ microorganisms = Sugar + Lipids + Amino\ acids \qquad (1.1)$$

$$Sugar + Lipids + Amino\ acids = CO_2 + H_2O + minerals \qquad (1.2)$$

In a case, microorganisms involved in nitrate reduction, aerobic degradation, and sulfate reduction were used to see the degradation of rhamnolipid and Triton X-100. It was observed that rhamnolipid degraded completely whereas a partial degradation of Triton X-100 was only moderately biodegradable under aerobic conditions [69]. BSs were found less toxic in comparison to different chemical-based surfactants while tested on invertebrates [70]. Various models of toxicity testing for the BSs have been established in the plant, microbial, and animals and their cell lines under laboratory conditions [58, 71, 72] (Table 1.3). It was Kanga et al. [85] who has established the fact that, glycolipids obtained from Rhodococcus species 413A were 50% less toxic than Plysorbate 80. All the reports and data obtained from different studies endorsed the fact that BSs are less toxic as equating to the synthetic surfactants. Structurally most of the BSs displayed significant EC_{50} concentration values than chemical-based dispersants [8].

1.5.3   Efficiency in Comparison with Chemical Surfactants

A comparison between BSs and their chemical counterparts is drawn to demonstrate variances in their structural belongings and advantages. Information about BSs properties systematized along these ranks is beneficial for those looking to formulate self-styled green or safe products with new perspectives. An assessment between BSs and synthetic surfactants displayed the unique contest of relating two types of molecules that eventually have the comparable function of being a surface-active molecule, but diverge quite expressively concerning their origins. All such differences will have a thoughtful influence on a diversity of properties like biode-gradability, environmental toxicity, chemical behavior, and cost. Regarding a solo approach for comparing BSs and chemical surfactants, would not be appropriate as no one tactic would consistently measure characteristics, usage, and disposal.

Biological surface-active agents or "biosurfactants" are amphiphile composed of the hydrophilic and hydrophobic tail. Chemically origin surfactants are largely petroleum-based. So, BSs have a clear advantage over synthetic surfactants, which showed lower toxicity, better degradability, and more ecological compatibility. Whereas synthetic surfactants are typically partially degradable or non-biodegradable so remain recalcitrant and toxic to the ecological niche. Chemical surfactants may gather and their production procedures and waste products can be harmful to the environment. Chemical-based surface-active agents can be petroleum or vegetable-based typically incorporated in cleaning, laundry, and personal care products. The major drawback is the exhaustion of a non-renewable resource and environmental pollution while producing petroleum-based surfactants.

Table 1.3 Different toxicity models assessment for biosurfactants reported

Biosurfactants	Phyto-toxicity	Animal model	Eco-toxicity	Cell lines	Reference
Glycolipid	Vegetable seeds	NP	Anomalocardia brasiliana, Artemia salina (larvae)	NP	Santos et al. [73]
Glycolipid	Brassica oleracea, Cichorium intybus, Solanum gilo	NP	NP	NP	de Souza Sobrinho et al. [74]
Mono-rhamnolipid	NP	Zebrafish	Microtox assay	(H1299) human lung cell line	Hogan et al. [75]
Lipopeptide	NP	Acute toxicity on mice	NP	NP	Anyanwu et al. [76]
Lipopeptides	NP	Acute toxicity on mice (LD$_{50}$)	NP	NP	Sahnoun et al. [77]
Lipopeptides	NP	NP	Artemia franciscan larvae	NP	Kiran et al. [57]
Lipopeptides & Rhamnolipids	NP	NP	Anopheles stephensi larvae	NP	Parthipan et al. [78]
Mono and di-rhamnolipid congeners	NP			Mouse L292 fibroblastic cell line	Patowary et al. [79]
Surfactin & Fengycin	NP	Daphnia similis	NP	NP	Trejo-Castillo et al. [80]
Lipopeptide	NP	Acute toxicity tests, carried out with Daphnia magna	NP	NP	Catter et al. [81]
Surfactin	NP	Acute dermal irritation, Acute oral toxicity	NP	NP	Fei et al. [82]
Glycolipid	NP	NP	Short-term acute toxicity test with Artemia nauplii	NP	Zenati et al. [83]
Glycolipid	Brassica nigra and Triticum aestivum	NP	NP	Mouse fibroblast (ATCC L929)	Sharma et al. [71]
Xylolipid	NP	NP	Oral and dermal toxicity in mice	NP	Saravanakumari and Mani [84]

Table 1.4 Comparison of microbial and synthetic surfactants (Adopted and modified from [86])

Surfactant property	Sophorolipids	Rhamnolipids	Surfactin	Alkyl polyglycoside
Biodegradable	√	√	√	√
Renewable	√	√	√	√
Foaming	×	√	√	√
pH stability	×[a]	×[a]	×[a]	√
Availability at large scale	√	√	×	√
Application in food & healthcare	√	√	√	√
HLB value	10–13[b]	10–12[b]	10–12[b]	~17[b]

[a]Ester-free forms are resistant to alkaline and, up to a point, acid hydrolysis
[b]Higher will be the HLB value, greater will be the detergency, dispersal efficiency and o/w emulsification

Alkyl chain of BSs is shorter as compared to the synthetic surfactants, i.e. chemical counterpart is composed of alkyl chain containing ten or more carbon units. The low carbon unit content of the alkyl chain encourages aqueous solubility in microbial surfactants. Microbial surfactants have curiously low critical micelle concentrations compared to chemical surfactants. Microbial surfactants are also effective in restricting the greenhouse gases' consequence by decreasing the emission of carbon dioxide. So, therefore, BSs can be labeled as "green" as of their degradability behavior, HLB scale, applications, and relative stability over a wide array of physicochemical properties and environmental (Table 1.4).

Furthermore, the reply to this present situation hinges on what we mean by "efficiency." For any surface-active agent "efficiency" can be demarcated in different ways, i.e. critical micelle concentration, surface tension, and emulsification potential (the lower the concentration, the improved the surfactants), decrease in air-water surface tension, emulsification (how efficiently they stabilize two different mixtures with diverse surface tension). It is tedious to find systematics comparison which consists of a broad spectrum of surfactants. Providentially, the majority of the CMC determination is observed in water, making it convenient to relate between surfactants in principle. It might be a point of argument that a well-accepted pH and a demarcated salt concentration would be beneficial for standardization drives, but in the nonappearance of unanimously agreed with ionic strength, water is a standard reference point. A comparison of the CMC and surface tension has been made for BSs and some of the chemical surfactants (Table 1.5).

BSs properties like self-assembly of amphiphiles, decrease of surface, and reduction in interfacial tension, emulsification potential, and adsorption make it a striking formulation substitute in various segments. Comparing the critical micelle concentration of a lipopeptide obtained from *B. subtilis* EG1, displayed the CMC was 96%, 83%, and 45% lesser than the chemical surfactants, i.e. alkylbenzene sulfonate, Glucopone 215, and Glucopone 650, respectively. It was observed that lipopeptide biosurfactant found more effective in decreasing surface tension as compared to chemical-based surfactants [87]. Additionally in some reports, BSs were found more

Table 1.5 Comparison of surface activities of biosurfactant and chemical surfactants

Surfactant	CMC	Surface tension mN/m
SDS	7 mM	38
Triton X-100	140 mg/L	30
Tween 20	70 mg/L	30
Tween 80	17 mg/L	42
Rhamnolipids	40–20 mg/L	27
Surfactin	10–30 mg/L	27
Sophorolipids	10 mg/L	33.5
Mannosylerythritol lipids (MEL)	15 mg/L	29.2

stable to the synthetic surfactants at elevated temperatures, extreme pH, and ionic strength [88, 89]. The critical micelle concentration, aggregation number, HLB, and cloud point increase the emulsifying potential and surface activity of BSs [90]. In a case, the micelle formation by a surfactin and sodium dodecylbenzene sulfonate (SDOBS) were observed. The mix of surfactin and SDOBS displayed reduced electrostatic repulsion and endorsed hydrophobic interaction among the molecules [91]. Jadav et al. [92] reported that the introduction of BSs into water improved the methane hydrate formation and decrease the induction time as equated to SDS. In another case, rhamnolipid has also exhibit higher emulsification activity with kerosene as compared to the SDS in the pH range 6–9 [91].

Various reports are available in which BSs toxicity has been explored in plants, aquatic life, animal, and human cell lines. Glycolipid obtained from *Candida lipolytica* was found non-toxic to seed germination and brine shrimp which showed no signs of cell toxicity [93]. In a recent observation, there was a comparison of toxicities among synthetic and biological rhamnolipid. Both the type of molecules displayed EC_{50} levels equivalent to a "somewhat toxic" environmental protection agency (EPA) rating for aquatic environments. Even, the data of biodegradability and cell toxicity in a human cell line hinge on the stereo-chemistry of the chemically synthesized rhamnolipid [75].

Emulsification potential is also measured by considering its HLB value, which entitles BSs is to be anticipated to form water-in-oil (w/o) or oil-in-water (o/w) emulsion. The HLB values have been analyzed and values fix on a gage of 0–20. BSs for suitable applications can be classified grounded on HLB values. Bio-emulsifiers with elevated HLB values confer significant solubility of oil-in-water. Typically, HLB values between 3 and 6, indulge in w/o microemulsions, whereas HLB values between 8 and 18 anticipate o/w microemulsions. For example, surfactin and rhamnolipids, conferring to their HLB gage, errand the development of o/w emulsions (Table 1.6) [94].

On the front of BSs ecological and environmental applications, some studies were observed that the effect of various non-ionic surfactants such as polysorbate-80, 60, 40, 20, and Triton X-100 and a glycolipid rhamnolipid on of Paenibacillus sp. PRNK-6 on the breakdown of fluorene. An enrichment in the growth and fluorene degradation was excluding polysorbate 20 and Triton X-100. Some of the

Table 1.6 HLB comparison of commonly known biosurfactants vs. chemical surfactants

Surfactant	HLB value
Oleic acid	1
Glycerol monostearate	11
Sorbitan monolaurate	9
Rhamnolipids	10.17
Sophorolipids	10–13
Surfactin	10–12
MEL	≥ 12
Other glycolipids	10–15
Lipopeptides	10–11.1
Polysorbate 80	14.4–15.6

surfactants such as Triton X-100 and polysorbate 20 were found toxic [95]. The maximal fluorene degradation rate was attained when polysorbate-80, 40, and rhamnolipid were utilized, while polysorbate-60 appears to be less active in the improvement of the fluorene degradation rate. Enhanced degradation is possibly due to the increase in cell surface hydrophobicity which rouses the uptake via direct interaction among cells and hydrocarbon molecules.

1.6   Surface and Interface Activity

When a surface-active agent is mixed, it will typically adsorb onto the surface of the solution and decrease the surface tension till the bulk concentration ranges to a point regarded as the critical micelle concentration (CMC). The surface tension of any suspension will not change practically above the CMC concentration. BSs are possibly suitable in terms of a decrease in surface tension as equate to the surface tension of the distill water, i.e. 72 mN/m. BSs surface tension was compared, depicted, and demonstrated in various studies (Table 1.7).

1.6.1   Emulsification and De-emulsification

An emulsion is defined as a mixture of two or more suspensions that are typically immiscible. Emulsions are part of an added all-purpose class of two-phase suspensions of matter known as colloids. The term emulsion and colloid are occasionally utilized interchangeably. Even though the terms colloid and emulsion are occasionally used interchangeably. Emulsification plays a vital part in the uniformity and texture of various food products. Emulsions are mostly of two types: oil-in-water (O/W), when the oil phase is dispersed whereas the aqueous phase is continuous; and water-in-oil (W/O), when the aqueous phase is dispersed in the oil phase [102]. Emulsions are thermodynamically uneven that lean towards to separate to their original state. To formulate industrial products with better shelf life, the addition of emulsifying agents is essential. Because of their potential to decrease

Table 1.7 Comparison of biosurfactants and chemicals surfactants in terms of surface tension reduction

Biosurfactants/surfactants	Surface tension (mN/m)	Reference
Biosurfactants		
Rhamnolipid	27 mN/m	Mendes et al. [96]
Surfactin	27 mN/m	Saharan et al. [2]
Sophorlipids	33 mN/m	Mulligan [97]
Lactobacillus fermentum RC-14 derived biosurfactants	39 mN/m	Velraeds et al. [98]
Streptococcus thermophilus B derived biosurfactants	37 mN/m	Velraeds et al. [99]
Streptococcus thermophilus A derived biosurfactants	36 mN/m	Rodrigues et al. [100]
Lactococcus lactis 53 derived biosurfactants	36 mN/m	Rodrigues et al. [101]
Chemical surfactants		
SDS	36 mN/m	Mendes et al. [96]

the interfacial tension among the continuous and dispersed phases, surfactants decrease the surface energy and the ratio of force needs to disperse one phase into the another, avoiding the coalescence of the molecules and the consequent phase separation [103]. HLB value is one vital constraint to determine the emulsification potential of any surfactant. HLB values vary flanked by 1 and 20. Surfactants with small HLB values (3–9) encourage the development of W/O emulsions, while HLB values more than 9 tend to stabilize O/W emulsions. For example, surfactin and rhamnolipids, conferring to their high HLB values, indulge in the formation of O/W emulsions [94].

Concerning the chemical surfactants utilized in food processing, regardless of their significant emulsifying potential and an extensive spectrum of applications, they are steadily bringing up the rear fondness, due to a growing demand between customers for natural substitutes to chemically derived additives. Moreover, regulatory guidelines and limits forced by food regulatory agencies and health establishments are continuously restricting their use [9, 104]. Approximately 500 thousand tons of surfactants are produced every year for food purposes, and surfactant demand in the European Union (EU) and the United States (US) are anticipated to be €300 million and US$275 million, correspondingly [105]. The present market epitomizes a prospect for new emergent emulsifiers, as well as microbial surfactants as the outcome of their effective surface and emulsifying potential.

Waste oil originated from the storage and transport industry during the refining process results in highly constant mixtures because of the natural occurrence of surfactant compounds such as resins, naphthenic acids, wax, and clay [106]. Such molecules are typically contained of 30–90% oil, 30–70% water, and 2–15% solids by weight and are present as a multifaceted type of water-in-oil (W/O) emulsion

[107, 108]. Such emulsions cannot be right disposed of to the environment because of high water composition and eco-toxicity [109]. So, demulsification is essential to disperse this remaining oil from the oil and water phases [110]. De-emulsification is attained by upsetting the thermodynamic circumstances at the interface. From a procedural point of view, oil industries are often concerned with three facets of demulsification:

- The rate at which the process of demulsification occurs
- Quality of demulsified water for consequent disposal
- The quantity of water residual in the oil after demulsification

In de-emulsification flocculation is the first phase, where microbial aggregate, forming flocs. Subsequently, in the second step, the large coalescence of oil droplets occurs [111]. The de-emulsification potential of a mixed bacterial population was typically analyzed by utilizing a petroleum-oil emulsion phase, about 96% of de-emulsification was attained. A mixed culture approach exhibits significant de-emulsifying potential as equated to the axenic or monoculture approach [112].

1.6.2 Foaming, Moisturizing, Dispersing, and Detergency Properties

Surface-active agents get accumulated at a gas–liquid interface resulting in the development of bubbles resulting in the foam formation. The examination of the foaming potential of surfactin seems predominantly attractive. As surfactin composed of both a lipidic chain, typically found in surfactant bio-molecules, and a peptidic chain. Surfactin is a hybrid structure that encouraged us to evaluate its capacity to form and stabilize foams. The foaming potential of the surfactin was compared with the SDS [113]. So, lipopeptide structures could help as a structural model to generate actual foaming agents, which are of significant attention for product development in detergent, cosmetic preparations, and pharmaceutical role.

Owing to diverse structural properties, biosurfactants with low toxicity, glycolipids can be used as next-generation additives in cosmetic formulations. Cosmetic moisturizers are agents particularly intended to make the external coatings of the skin softer and more pliant. Moisturizers upsurge the skin's water content or hydration by decreasing evaporation and display skin and hair care effectiveness [114].

Glycolipids can be used as moisturizers letting their use in cosmetics [115]. Dispersion of colloidal particles to attain a stable phase is a subject of great importance for food, paint, pharmaceutical, and cosmetics. A dispersant agent is a kind of molecule that decreases the cohesive hold among similar particles. Such potential of the surfactant keeps immiscible particles in suspension by avoiding insoluble to from coalescence aggregations of the formulation. BSs would improve the aqueous dispersion of organic substances having limited water solubility in two behaviors:

1. BSs would reduce the surface and interfacial tensions in the production medium, which ultimately increasing the aqueous phase dispersion

2. The potential of a rhamnolipid to raise the aqueous dispersion of slightly soluble organic substances and to improve the rate of biodegradation.

The rhamnolipid biosurfactant has been used for the bioremediation in a batch set-up as a potential dispersion agent, increasing the degradation of octadecane 4 times in a pure-culture batch experiment [116]. It was found that the rhamnolipids can increase octadecane dispersion to 75 mg/L at static conditions. In another report, rhamnolipid was found effective dispersion of colloidal alumina.

Surface-active agents play a crucial role in laundry and household detergent products. The biological origin with less toxicity and degradability leads us to accomplish that the future use of BSs as laundry detergent additive is extremely promising. The utilization of the environmentally friendly surfactants obtained from cost-effective resources in the detergent formulation is a developing inclination.

1.7   Surfactant Vs. Biosurfactants

Surfactants stand for "surface-active agents" are amphiphilic molecules that reduce the surface and interfacial tension among two surfaces. BSs are amphiphilic molecules composed of lipophilic and hydrophilic moieties that make them appropriate for cleaning and surface properties. The surfactant is the vital active fraction in most cleaning products and aids two key roles: First, it aids to "wet" the surface, secondly, surfactant emulsifies oils and fats deposition. The majority of the synthetic surfactants are originated from petroleum or vegetable-based feedstock. The disadvantage of utilizing petroleum-derived surfactants is that they lead to the exhaustion of a non-renewable reserve and are extremely polluting. They are only biodegradable to some extent and accumulates recalcitrant chemicals in the environment. Subsequently, synthetic surfactants can cause skin irritation and internal allergies in certain cases. Most of the synthetic surfactants are composed of chemical compounds, which in some scenarios can produce irritant reactions. Synthetic surfactants can interact with epidermal lipids and proteins, which leads to the removal of fatty acids by distorting the intercellular arrangements and also disturb the skin cells [117].

In the last decade, the demand for natural, non-toxic, green-labeled formulations obtained from microbial sources has extensively increased in cosmetic and pharmaceutical formulations. In such a direction, more biocompatible surfactants with lower toxic effects than their synthetic counterparts are being seen for. Recently, it has been observed that BSs can progress the preservation and active fraction solubility utilized in food and pharmaceutical formulations, as an effective alternative to synthetic ones, which can root important damages. In comparison with synthetic surfactants, BSs are more biodegradable due to their macromolecular structure composed of lipids, polysaccharides, proteins, and peptides.

BSs are obtained from 100% natural and biodegradable produced by microorganisms. Similar to the plant and petroleum-based surfactants, they have potential emulsifying, foam forming, wetting, or even dispersing characteristics. BSs also have efficient surface tension-reducing capacity which makes it convenient to get rid of oils and fats in a small concentration.

The strong current desire for a "greener molecule" for industrial sustainability has encouraged active interest in microbial surfactants as a possible substitute for dozens of applications [6].

1.7.1 Competence of Biosurfactants in Comparison with Chemical Surfactants

The competence of microbial surfactants towards chemical surfactants has many aspects to compare. Some of them have been reviewed and compared with quantifiable parameters. In one experimental demonstration, rhamnolipids derived from the *P. aeruginosa* were observed to be more competent than chemical surfactants such as polysorbate 60, SDS, polyoxyethylene, and sorbitan monooleate. Similarly, BSs obtained *Candida ishiwadae*, a MEL BSs produced by *Candida antarctica,* and glycolipid obtained from *Trichosporon asahii* displayed better surfactant potential than various synthetic surfactants. Glycolipidic BSs produced by the *Bacillus methylotrophicus* were found to be better than SDS in the emulsification of hydrocarbon emulsification [118].

BSs exhibit significant emulsification properties as compared to the synthetic counterpart. BSs produced by the *P. aeruginosa* were found significantly better emulsifying agents against different oils such as motor, peanut, and xylene, naphthalene, and anthracene. Similar experiments were planned with a chemical counterpart, i.e. as Triton X-100 at 1 mg/mL [119]. As a glycolipid, rhamnolipid outpace non-ionic synthetic surfactants at the same concentration and also out pass SDS. In a case, rhamnolipids were able to solubilize organic dye such as Sudan III various fold as compared to the SDS. Rhamnolipids are known for the superior colloidal properties as compared to synthetic surfactants. On the other hand, surfactin is also competent as the concentration required to achieve CMC or micellization is extremely low which leads to early aggregation. A low CMC also anticipates a low degree of exchange with both solution and the surface, which ultimately can stabilize emulsions potentially.

1.8 Aptness of Biosurfactants for Industrial and Environmental Applications

The utilization of BSs in industrial and biomedical applications is attaining new heights because of various physicochemical properties (Table 1.8). Microbial enhanced oil recovery, solubilization of polyaromatic hydrocarbon, emulsification, dispersing agents are some of the well-studied mechanisms. Food applications, biomedical utilization, pharmaceutical adoption, and a new range of cleaning materials would be future applications. Some of those prominent properties have

Table 1.8 Biosurfactants properties and their aptness for industrial applications

Properties of biosurfactants	Possible applications	Prospects
Biodegradable	Environmental pollution control of PAH, PAC, and heavy metals	Biosorbent molecules
Non-toxic	Food, pharmaceuticals, healthcare	Drug delivery, cleaning agents
Specificity	Antimicrobial, anti-protozoan, Mosquitocidal, antifungal	Insect control, alternative antibiotics
Thermal stable	Food processing and microbial enhanced oil recovery	Bakery and detergents
pH stable	Waste management, soil washing, food processing	Acidic food emulsification and preservation
Salt stability	Environmental, marine application	Detergent, high salt food
Surface active	Modification of gas, liquid interfaces	Mixing of hydrophobic and hydrophilic suspension
Emulsification	Oil recovery, food emulsions	Cosmetics and bakery applications
Ionic strength	Removal of heavy metals	Washing of food surfaces
Compatibility	Direct application to human, plant, and animal health	New frontiers in food, feed, and plant growth
Wettability	Oil spill control, cleaning of aquifers	In situ removal of oil from industrial sites
Dispersion	Paint and pesticide	Specific pesticide
Detergency	Cosmetic and healthcare	Oral and personal hygiene
Antimicrobial	Food and cosmetics	Conjugates, synergistic with known antimicrobials
Antibiofilm	Control of biofilm	Biofilm control on food and industrial surfaces
Antifungal	Ointments and antifungal suspensions	Tropical infections treatment

been summarized which makes BSS more apt towards industrial and environmental applications.

1.9 Concluding Remark

Even though BSs have been considered as a promising substitute for synthetic surfactants for years, but they still need to deliver on their efficacy on various fronts. It is not only the scientific constraints but, certain cost factors are also responsible for the underutilization of the BSs, not the scientific constraints. There should be a serious comparison of the physical-chemical properties and subsequent applications. It is hopeful to see that there is still abundant space is available for basic research of the molecular interactions of their surface potential, and it will be a huge step frontward to be able to fit this information into a better inclusive considerate of the interactions among microorganisms and their applications in biosurfactants applications as a greener sustainable agents.

References

1. Sharma D, Saharan BS, Kapil S (2016) Biosurfactants of lactic acid bacteria. Springer, Cham
2. Saharan BS, Sahu RK, Sharma D (2011) A review on biosurfactants: fermentation, current developments and perspectives. Genet Eng Biotechnol J 2011(1):1–14
3. Satpute SK, Mone NS, Das P, Banat IM, Banpurkar AG (2019) Inhibition of pathogenic bacterial biofilms on PDMS based implants by L. acidophilus derived biosurfactant. BMC Microbiol 19(1):1–15
4. Kumar P, Sharma PK, Sharma PK, Sharma D (2015) Micro-algal lipids: a potential source of biodiesel. JIPBS 2(2):135–143
5. Sharma D, Dhanjal DS, Mittal B (2017) Development of edible biofilm containing cinnamon to control food-borne pathogen. J Appl Pharm Sci 7(01):160–164
6. Banat IM, Franzetti A, Gandolfi I, Bestetti G, Martinotti MG, Fracchia L et al (2010) Microbial biosurfactants production, applications and future potential. Appl Microbiol Biotechnol 87 (2):427–444
7. Bodour AA, Drees KP, Maier RM (2003) Distribution of biosurfactant-producing bacteria in undisturbed and contaminated arid southwestern soils. Appl Environ Microbiol 69:3280–3287
8. Desai JD, Banat IM (1997) Microbial production of surfactants and their commercial potential. Microbiol Mol Biol Rev 61:47–64
9. Nitschke M, Costa SGVAO (2007) Biosurfactants in food industry. Trends Food Sci Technol 18(5):252–259
10. Sharma D, Saharan BS (eds) (2018) Microbial cell factories. CRC Press, Boca Raton
11. Rodrigues LR, Teixeira JA (2010) Biomedical and therapeutic applications of biosurfactants. In: Biosurfactants. Springer, New York, pp 75–87
12. Saha P, Nath D, Choudhury MD, Talukdar AD (2018) Probiotic biosurfactants: a potential therapeutic exercises in biomedical sciences. In: Microbial biotechnology. Springer, Singapore, pp 499–514
13. Batista SB, Mounteer AH, Amorim FR, Totola MR (2006) Isolation and characterization of biosurfactant/bioemulsifier-producing bacteria from petroleum contaminated sites. Bioresour Technol 97(6):868–875
14. Kuyukina MS, Ivshina IB, Makarov SO, Litvinenko LV, Cunningham CJ, Philp JC (2005) Effect of biosurfactants on crude oil desorption and mobilization in a soil system. Environ Int 31(2):155–161
15. Sekhon Randhawa KK, Rahman PK (2014) Rhamnolipid biosurfactants—past, present, and future scenario of global market. Front Microbiol 5:454
16. Mukherjee S, Das P, Sen R (2006) Towards commercial production of microbial surfactants. Trends Biotechnol 24(11):509–515
17. Sharma D, Saharan BS (2016) Functional characterization of biomedical potential of biosurfactant produced by Lactobacillus helveticus. Biotechnol Rep 11:27–35
18. Otzen DE (2017) Biosurfactants and surfactants interacting with membranes and proteins: same but different? Biochimica et Biophysica Acta (BBA)-Biomembranes 1859(4):639–649
19. Dembitsky VM (2004) Astonishing diversity of natural surfactants: 1. Glycosides of fatty acids and alcohols. Lipids 39(10):933–953
20. Varjani SJ, Upasani VN (2016) Carbon spectrum utilization by an indigenous strain of Pseudomonas aeruginosa NCIM 5514: production, characterization and surface active properties of biosurfactant. Bioresour Technol 221:510–516
21. Bergström S, Theorell H, Davide H (1946) On a metabolic product of Ps. Pyocyanea, pyolipic acid, active against Mycobacterium tuberculosis. Ark Chem Miner Geol 23A(13):1–12
22. Burger MM, Glaser L, Burton RM (1963) The enzymatic synthesis of a rhamnose-containing glycolipid by extracts of Pseudomonas aeruginosa. J Biol Chem 238(8):2595–2602
23. Dubeau D, Déziel E, Woods DE, Lépine F (2009) Burkholderia thailandensis harbors two identical rhl gene clusters responsible for the biosynthesis of rhamnolipids. BMC Microbiol 9 (1):263

24. Soberón-Chávez G, Lépine F, Déziel E (2005) Production of rhamnolipids by Pseudomonas aeruginosa. Appl Microbiol Biotechnol 68(6):718–725

25. Li Q (2017) Rhamnolipid synthesis and production with diverse resources. Front Chem Sci Eng 11(1):27–36

26. Maier RM, Soberon-Chavez G (2000) Pseudomonas aeruginosa rhamnolipids: biosynthesis and potential applications. Appl Microbiol Biotechnol 54(5):625–633

27. Zhu K, Rock CO (2008) RhlA converts β-hydroxyacyl-acyl carrier protein intermediates in fatty acid synthesis to the β-hydroxydecanoyl-β-hydroxydecanoate component of rhamnolipids in Pseudomonas aeruginosa. J Bacteriol 190(9):3147–3154

28. Rahim R, Ochsner UA, Olvera C, Graninger M, Messner P, Lam JS, Soberón-Chávez G (2001) Cloning and functional characterization of the Pseudomonas aeruginosa rhlC gene that encodes rhamnosyltransferase 2, an enzyme responsible for di-rhamnolipid biosynthesis. Mol Microbiol 40(3):708–718

29. Pearson JP, Pesci EC, Iglewski BH (1997) Roles of Pseudomonas aeruginosa las and rhl quorum-sensing systems in control of elastase and rhamnolipid biosynthesis genes. J Bacteriol 179(18):5756–5767

30. Rooney AP, Price NP, Ray KJ, Kuo TM (2009) Isolation and characterization of rhamnolipid-producing bacterial strains from a biodiesel facility. FEMS Microbiol Lett 295(1):82–87

31. Nayak AS, Vijaykumar MH, Karegoudar TB (2009) Characterization of biosurfactant produced by Pseudoxanthomonas sp. PNK-04 and its application in bioremediation. Int Biodeterior Biodegradation 63(1):73–79

32. Gorin PAJ, Spencer JFT, Tulloch AP (1961) Hydroxy fatty acid glycosides of sophorose from Torulopsis magnoliae. Can J Chem 39(4):846–855

33. Tulloch AP, Spencer JFT, Gorin PAJ (1962) The fermentation of long-chain compounds by torulopsis magnoliae: I. structures of the hydroxy fatty acids obtained by the fermentation of fatty acids and hydrocarbons. Can J Chem 40(7):1326–1338

34. Van Bogaert IN, Zhang J, Soetaert W (2011) Microbial synthesis of sophorolipids. Process Biochem 46(4):821–833

35. Anderson RJ, Newman MS (1933) The chemistry of the lipids of tubercle bacilli XXXIII. Isolation of trehalose from the acetone-soluble fat of the human tubercle bacillus. J Biol Chem 101(2):499–504

36. Franzetti A, Gandolfi I, Bestetti G, Smyth TJ, Banat IM (2010) Production and applications of trehalose lipid biosurfactants. Eur J Lipid Sci Technol 112(6):617–627

37. Fukuoka T, Morita T, Konishi M, Imura T, Sakai H, Kitamoto D (2007) Structural characterization and surface-active properties of a new glycolipid biosurfactant, mono-acylated mannosylerythritol lipid, produced from glucose by Pseudozyma Antarctica. Appl Microbiol Biotechnol 76(4):801–810

38. Fan L, Xie P, Wang Y, Huang Z, Zhou J (2018) Biosurfactant–protein interaction: influences of mannosylerythritol lipids-a on β-glucosidase. J Agric Food Chem 66(1):238–246

39. Kitamoto D, Akiba S, Hioki C, Tabuchi T (1990) Extracellular accumulation of mannosylerythritol lipids by a strain of Candida antarctica. Agric Biol Chem 54(1):31–36

40. Singh P, Cameotra SS (2004) Potential applications of microbial surfactants in biomedical sciences. Trends Biotechnol 22(3):142–146

41. Gong Z, Peng Y, Zhang Y, Song G, Chen W, Jia S, Wang Q (2015) Construction and optimization of Escherichia coli for producing rhamnolipid biosurfactant. Sheng wu gong cheng xue bao= Chinese J Biotechnol 31(7):1050–1062

42. Moyne AL, Shelby R, Cleveland TE, Tuzun S (2001) Bacillomycin D: an iturin with antifungal activity against Aspergillus flavus. J Appl Microbiol 90(4):622–629

43. Vanittanakom N, Loeffler W, Koch U, Jung G (1986) Fengycin-a novel antifungal lipopeptide antibiotic produced by Bacillus subtilis F-29-3. J Antibiot 39(7):888–901

44. Peypoux F, Guinand M, Michel G, Delcambe L, Das BC, Lederer E (1978) Structure of iturine a, a peptidolipid antibiotic from Bacillus subtilis. Biochemistry 17(19):3992–3996

45. Thimon L, Peypoux F, Wallach J, Michel G (1995) Effect of the lipopeptide antibiotic, iturin a, on morphology and membrane ultrastructure of yeast cells. FEMS Microbiol Lett 128 (2):101–106

46. Moyne AL, Cleveland TE, Tuzun S (2004) Molecular characterization and analysis of the operon encoding the antifungal lipopeptide bacillomycin D. FEMS Microbiol Lett 234 (1):43–49

47. Tsuge K, Ohata Y, Shoda M (2001) Gene yerP, involved in surfactin self-resistance in Bacillus subtilis. Antimicrob Agents Chemother 45(12):3566–3573

48. Rosenberg E, Ron EZ (1999) High-and low-molecular-mass microbial surfactants. Appl Microbiol Biotechnol 52(2):154–162

49. Kaplan N, Rosenberg E, Jann B, Jann K (1985) Structural studies of the capsular polysaccharide of Acinetobacter calcoaceticus BD4. Eur J Biochem 152(2):453–458

50. Navon-Venezia S, Zosim Z, Gottlieb A, Legmann R, Carmeli S, Ron EZ, Rosenberg E (1995) Alasan, a new bioemulsifier from Acinetobacter radioresistens. Appl Environ Microbiol 61 (9):3240–3244

51. Toren A, Orr E, Paitan Y, Ron EZ, Rosenberg E (2002) The active component of the bioemulsifier alasan from Acinetobacter radioresistens KA53 is an OmpA-like protein. J Bacteriol 184(1):165–170

52. Gutierrez T, Shimmield T, Haidon C, Black K, Green DH (2008) Emulsifying and metal ion binding activity of a glycoprotein exopolymer produced by Pseudoalteromonas sp. strain TG12. Appl Environ Microbiol 74(15):4867–4876

53. Linder MB (2009) Hydrophobins: proteins that self assemble at interfaces. Curr Opin Colloid Interface Sci 5(14):356–363

54. Saranya P, Swarnalatha S, Sekaran G (2014) Lipoprotein biosurfactant production from an extreme acidophile using fish oil and its immobilization in nanoporous activated carbon for the removal of ca 2+ and Cr 3+ in aqueous solution. RSC Adv 4(64):34144–34155

55. Wösten HA (2001) Hydrophobins: multipurpose proteins. Ann Rev Microbiol 55(1):625–646

56. Ram H, Sahu AK, Said MS, Banpurkar AG, Gajbhiye JM, Dastager SG (2019) A novel fatty alkene from marine bacteria: a thermo stable biosurfactant and its applications. J Hazard Mater 380:120868

57. Kiran GS, Priyadharsini S, Sajayan A, Priyadharsini GB, Poulose N, Selvin J (2017) Production of lipopeptide biosurfactant by a marine Nesterenkonia sp. and its application in food industry. Front Microbiol 8:1138

58. Sharma D, Saharan BS, Chauhan N, Procha S, Lal S (2015) Isolation and functional characterization of novel biosurfactant produced by Enterococcus faecium. Springerplus 4(1):1–14

59. Hmidet N, Jemil N, Nasri M (2019) Simultaneous production of alkaline amylase and biosurfactant by Bacillus methylotrophicus DCS1: application as detergent additive. Biodegradation 30(4):247–258

60. Derguine-Mecheri L, Kebbouche-Gana S, Khemili-Talbi S, Djenane D (2018) Screening and biosurfactant/bioemulsifier production from a high-salt-tolerant halophilic Cryptococcus strain YLF isolated from crude oil. J Pet Sci Eng 162:712–724

61. Bai L, McClements DJ (2016) Formation and stabilization of nanoemulsions using biosurfactants: Rhamnolipids. J Colloid Interface Sci 479:71–79

62. Kitamoto D, Yanagishita H, Endo A, Nakaiwa M, Nakane T, Akiya T (2001) Remarkable antiagglomeration effect of a yeast biosurfactant, diacylmannosylerythritol, on ice-water slurry for cold thermal storage. Biotechnol Prog 17(2):362–365

63. Saimmai A, Rukadee O, Onlamool T, Sobhon V, Maneerat S (2012) Isolation and functional characterization of a biosurfactant produced by a new and promising strain of Oleomonas sagaranensis AT18. World J Microbiol Biotechnol 28(10):2973–2986

64. Sharma V, Garg M, Devismita T, Thakur P, Henkel M, Kumar G (2018) Preservation of microbial spoilage of food by biosurfactantbased coating. Asian J Pharm Clin Res *11*(2):98

65. Singh J, Sharma D, Kumar G, Sharma, N. R. (Eds.). (2018) Microbial bioprospecting for sustainable development. Springer, Berlin

66. Muthusamy K, Gopalakrishnan S, Ravi TK, Sivachidambaram P (2008) Biosurfactants: properties, commercial production and application. Curr Sci 94:736–747
67. Berna JL, Cassani G, Hager CD, Rehman N, López I, Schowanek D et al (2007) Anaerobic biodegradation of surfactants–scientific review. Tenside Surfactants Detergents 44 (6):312–347
68. Garcia MT, Campos E, Dalmau M, Illan P, Sanchez-Leal J (2006) Inhibition of biogas production by alkyl benzene sulfonates (LAS) in a screening test for anaerobic biodegradability. Biodegradation 17(1):39–46
69. Mohan PK, Nakhla G, Yanful EK (2006) Biokinetics of biodegradation of surfactants under aerobic, anoxic and anaerobic conditions. Water Res 40(3):533–540
70. Edwards KR, Lepo JE, Lewis MA (2003) Toxicity comparison of biosurfactants and synthetic surfactants used in oil spill remediation to two estuarine species. Mar Pollut Bull 46 (10):1309–1316
71. Sharma D, Saharan BS, Chauhan N, Bansal A, Procha S (2014) Production and structural characterization of Lactobacillus helveticus derived biosurfactant. Sci World J:2014
72. Cochis A, Fracchia L, Martinotti MG, Rimondini L (2012) Biosurfactants prevent in vitro Candida albicans biofilm formation on resins and silicon materials for prosthetic devices. Oral Surg Oral Med Oral Pathol Oral Radiol 113(6):755–761
73. Santos DKF, Rufino RD, Luna JM, Santos VA, Sarubbo LA (2016) Biosurfactants: multi-functional biomolecules of the 21st century. Int J Mol Sci 17(3):401
74. de Souza Sobrinho B, de Souza Sobrinho HB, de Luna JM, Rufino RD, Figueiredo AL, Sarubbo PA (2013) Application of biosurfactant from Candida sphaerica UCP 0995 in removal of petroleum derivative from soil and sea water. J Life Sci 7(6):559–569
75. Hogan DE, Tian F, Malm SW, Olivares C, Pacheco RP, Simonich MT et al (2019) Biodegradability and toxicity of monorhamnolipid biosurfactant diastereomers. J Hazard Mater 364:600–607
76. Anyanwu CU, Obi SKC, Okolo BN (2011) Lipopeptide biosurfactant production by Serratia marcescens NSK-1 strain isolated from petroleum-contaminated soil. J Appl Sci Res 7 (1):79–87
77. Sahnoun R, Mnif I, Fetoui H, Gdoura R, Chaabouni K, Makni-Ayadi F et al (2014) Evaluation of Bacillus subtilis SPB1 lipopeptide biosurfactant toxicity towards mice. Int J Pept Res Ther 20(3):333–340
78. Parthipan P, Preetham E, Machuca LL, Rahman PK, Murugan K, Rajasekar A (2017) Biosurfactant and degradative enzymes mediated crude oil degradation by bacterium Bacillus subtilis A1. Front Microbiol 8:193
79. Patowary K, Patowary R, Kalita MC, Deka S (2017) Characterization of biosurfactant produced during degradation of hydrocarbons using crude oil as sole source of carbon. Front Microbiol 8:279
80. Trejo-Castillo R, Martínez-Trujillo MA, García-Rivero M (2014) Effectiveness of crude biosurfactant mixture for enhanced biodegradation of hydrocarbon contaminated soil in slurry reactor. International Journal of Environmental Research 8(3):727–732
81. Catter KM, Oliveira DFD, Sousa OVD, Gonçalves LRB, Vieira RHSDF, Alves CR (2016) Biosurfactant production by pseudomonas aeruginosa and burkholderia gladioli isolated from mangrove sediments using alternative substrates. Orbital: The Electronic Journal of Chemistry 8(5):269–275
82. Fei D, Zhou GW, Yu ZQ, Gang HZ, Liu JF, Yang SZ et al (2020) Low-toxic and nonirritant biosurfactant Surfactin and its performances in detergent formulations. J Surfactant Deterg 23 (1):109–118
83. Zenati B, Chebbi A, Badis A, Eddouaouda K, Boutoumi H, El Hattab M et al (2018) A non-toxic microbial surfactant from Marinobacter hydrocarbonoclasticus SdK644 for crude oil solubilization enhancement. Ecotoxicol Environ Saf 154:100–107

84. Saravanakumari P, Mani K (2010) Structural characterization of a novel xylolipid biosurfactant from Lactococcus lactis and analysis of antibacterial activity against multi-drug resistant pathogens. Bioresour Technol 101(22):8851–8854

85. Kanga SA, Bonner JS, Page CA, Mills MA, Autenrieth RL (1997) Solubilization of naphthalene and methyl-substituted naphthalenes from crude oil using biosurfactants. Environ Sci Technol 31(2):556–561

86. Fleurackers SJJ, Van Bogaert INA, Develter D (2010) On the production and identification of medium-chained sophorolipids. Eur J Lipid Sci Technol 112:655–662

87. Vaz DA, Gudina EJ, Alameda EJ, Teixeira JA, Rodrigues LR (2012) Performance of a biosurfactant produced by a Bacillus subtilis strain isolated from crude oil samples as compared to commercial chemical surfactants. Colloids Surf B: Biointerfaces 89:167–174

88. Penfold J, Chen M, Thomas RK, Dong C, Smyth TJ, Perfumo A et al (2011) Solution self-assembly of the sophorolipid biosurfactant and its mixture with anionic surfactant sodium dodecyl benzene sulfonate. Langmuir 27(14):8867–8877

89. Pereira JF, Gudiña EJ, Costa R, Vitorino R, Teixeira JA, Coutinho JA, Rodrigues LR (2013) Optimization and characterization of biosurfactant production by Bacillus subtilis isolates towards microbial enhanced oil recovery applications. Fuel 111:259–268

90. Jimenez Islas D, Medina Moreno SA, Gracida Rodriguez JN (2010) Biosurfactant properties, applications and production a review. Revista Internacional De Contaminacion Ambiental 26 (1):65–84

91. Lovaglio RB, dos Santos FJ, Junior MJ, Contiero J (2011) Rhamnolipid emulsifying activity and emulsion stability: pH rules. Colloids Surf B: Biointerfaces 85(2):301–305

92. Jadav S, Sakthipriya N, Doble M, Sangwai JS (2017) Effect of biosurfactants produced by Bacillus subtilis and Pseudomonas aeruginosa on the formation kinetics of methane hydrates. J Nat Gas Sci Eng 43:156–166

93. Santos DKF, Meira HM, Rufino RD, Luna JM, Sarubbo LA (2017) Biosurfactant production from Candida lipolytica in bioreactor and evaluation of its toxicity for application as a bioremediation agent. Process Biochem 54:20–27

94. Gudiña EJ, Rangarajan V, Sen R, Rodrigues LR (2013) Potential therapeutic applications of biosurfactants. Trends Pharmacol Sci 34(12):667–675

95. Reddy PV, Karegoudar TB, Nayak AS (2018) Enhanced utilization of fluorene by Paenibacillus sp. PRNK-6: effect of rhamnolipid biosurfactant and synthetic surfactants. Ecotoxicol Environ Saf 151:206–211

96. Mendes AN, Filgueiras LA, Pinto JC, Nele M (2015) Physicochemical properties of rhamnolipid biosurfactant from Pseudomonas aeruginosa PA1 to applications in microemulsions. J Biomater Nanobiotechnol 6(01):64

97. Mulligan CN (2005) Environmental applications for biosurfactants. Environ Pollut 133 (2):183–198

98. Velraeds MM, Van der Mei HC, Reid G, Busscher HJ (1996) Inhibition of initial adhesion of uropathogenic Enterococcus faecalis by biosurfactants from Lactobacillus isolates. Appl Environ Microbiol 62(6):1958–1963

99. Velraeds MM, Van Der Mei HC, Reid G, Busscher HJ (1997) Inhibition of initial adhesion of uropathogenic Enterococcus faecalis to solid substrata by an adsorbed biosurfactant layer from Lactobacillus acidophilus. Urology 49(5):790–794

100. Rodrigues LR, Teixeira JA, van der Mei HC, Oliveira R (2006a) Physicochemical and functional characterization of a biosurfactant produced by Lactococcus lactis 53. Colloids Surf B: Biointerfaces 49(1):79–86

101. Rodrigues L, Banat IM, Teixeira J, Oliveira R (2006b) Biosurfactants: potential applications in medicine. J Antimicrob Chemother 57(4):609–618

102. McClements DJ, Bai L, Chung C (2017) Recent advances in the utilization of natural emulsifiers to form and stabilize emulsions. Ann Rev Food Sci Technol 8:205–236

103. Kralova I, Sjöblom J (2009) Surfactants used in food industry: a review. J Dispers Sci Technol 30(9):1363–1383

104. Luna JM, Rufino RD, Sarubbo LA, Campos-Takaki GM (2013) Characterisation, surface properties and biological activity of a biosurfactant produced from industrial waste by Candida sphaerica UCP0995 for application in the petroleum industry. Colloids Surf B: Biointerfaces 102:202–209

105. Hasenhuettl GL, Hartel RW (eds) (2008) Food emulsifiers and their applications (Vol. 40, no. 6). Springer, New York

106. Jiang L, Yang HQ, Tao YR (2011) Biosurfactants and its application in biodegradation [J]. Hubei Agricul Sci 17

107. Yang H, Irudayaraj J, Paradkar MM (2005) Discriminant analysis of edible oils and fats by FTIR, FT-NIR and FT-Raman spectroscopy. Food Chem 93(1):25–32

108. Zhang X, Xu D, Zhu C, Lundaa T, Scherr KE (2012) Isolation and identification of biosurfactant producing and crude oil degrading Pseudomonas aeruginosa strains. Chem Eng J 209:138–146

109. Cambiella A, Benito JM, Pazos C, Coca J (2006) Centrifugal separation efficiency in the treatment of waste emulsified oils. Chem Eng Res Des 84(1):69–76

110. She YH, Zhang F, Xia JJ, Kong SQ, Wang ZL, Shu FC, Hu JM (2011) Investigation of biosurfactant-producing indigenous microorganisms that enhance residue oil recovery in an oil reservoir after polymer flooding. Appl Biochem Biotechnol 163(2):223–234

111. Kokal SL (2005) Crude oil emulsions: a state-of-the-art review. SPE Prod facil 20(01):5–13

112. Nadarajah N, Singh A, Ward OP (2002) De-emulsification of petroleum oil emulsion by a mixed bacterial culture. Process Biochem 37(10):1135–1141

113. Razafindralambo H, Paquot M, Baniel A, Popineau Y, Hbid C, Jacques P, Thonart P (1996) Foaming properties of surfactin, a lipopeptide biosurfactant from Bacillus subtilis. J Am Oil Chem Soc 73(1):149–151

114. Morita T, Kitagawa M, Suzuki M, Yamamoto S, Sogabe A, Yanagidani S et al (2009) A yeast glycolipid biosurfactant, mannosylerythritol lipid, shows potential moisturizing activity toward cultured human skin cells: the recovery effect of MEL-A on the SDS-damaged human skin cells. J Oleo Sci 58(12):639–642

115. Morita T, Koike H, Koyama Y, Hagiwara H, Ito E, Fukuoka T et al (2013) Genome sequence of the basidiomycetous yeast Pseudozyma Antarctica T-34, a producer of the glycolipid biosurfactants mannosylerythritol lipids. Genome Announ 1(2)

116. Zhang YIMIN, Miller RM (1992) Enhanced octadecane dispersion and biodegradation by a pseudomonas rhamnolipid surfactant (biosurfactant). Appl Environ Microbiol 58 (10):3276–3282

117. Vecino X, Cruz JM, Moldes AB, Rodrigues LR (2017) Biosurfactants in cosmetic formulations: trends and challenges. Crit Rev Biotechnol 37(7):911–923

118. Chandankere R, Yao J, Cai M, Masakorala K, Jain AK, Choi MM (2014) Properties and characterization of biosurfactant in crude oil biodegradation by bacterium Bacillus methylotrophicus USTBa. Fuel 122:140–148

119. Thavasi R, Nambaru VS, Jayalakshmi S, Balasubramanian T, Banat IM (2011) Biosurfactant production by Pseudomonas aeruginosa from renewable resources. Indian J Microbiol 51 (1):30–36

Screening of Biosurfactants

2

Abstract

The development of various screening approaches which are rapid, as well as reliable, and selection of a population of microorganisms as efficient BSs producer followed by a successive assessment of surface activity, holds the key for the identification of new biosurfactants. Screening for new and more potent biological surfactants continue, as microbial surfactants are well recognized for antimicrobial, antiviral, biomedical, and environmental applications. Though, the application of BSs is restricted due to the low product yield by many known strains and the cost of production medium are also one of the major hindrances. Hence, the search for new BS producers from different environmental sites with improved and novel properties is an ongoing attempt. With the increasing demand and interest in BSs, different approaches for identifying potential strains have been developed. But, the existing methods have some drawbacks like they are time-consuming, involve sophisticated equipment and reagents, or complicated preparation. The present chapter targets the various known methods along with the comparison of their advantages and drawbacks.

Keywords

Screening of BSs · Qualitative and Quantitative screening · High-throughput screening · Metagenomics

2.1 Properties of Biosurfactants or Basis of Screening

Emulsion If an immiscible solution such as nonpolar hydrocarbon is mixed with an aqueous solution in the presence of surfactant, the hydrophilic and hydrophobic moieties of surfactant remain orient itself towards the aqueous and hydrocarbon phase, respectively. The dispersion of water and hydrocarbon in one another is known as an emulsion [1–4].

© Springer Nature Singapore Pte Ltd. 2021
D. Sharma, *Biosurfactants: Greener Surface Active Agents for Sustainable Future*,
https://doi.org/10.1007/978-981-16-2705-7_2

Emulsification Emulsification is a process of formation of emulsion, composed of very small droplets non-coalescence of fat dispersed in water [5]. Emulsifiers are generally used to apply as an additive to enhance bioremediation and exclusion of hydrocarbons from environment [5, 6]. The emulsion forming potential of an amphiphilic molecule can be used as the BSs screening tool.

Micelle Formation A micelle is an aggregate of surfactant molecules present in a liquid, developing a colloidal suspension. With an increase in surfactant concentration, a critical concentration is attained after which there is no change in surface and interfacial characteristics [4, 7]. In micellar arrangements, hydrophobic "tails" of surface-active agents are aligned towards an oil droplet and hydrophilic "heads" inclined towards the aqueous phase [1–3]. The alignment of microbial molecules to form micelles can be used as an effective criterion to screen the large microbial population for BSs production.

Reduction of Surface and Interfacial Tensions BSs tend to accumulate at the interface between two immiscible liquids. The role of the BSs is to reduce the repulsive forces among diverse phases which ultimately allow them to interact more efficiently [8]. Direct measurement of the reduction in surface and interfacial using high-end tensiometers and ADSA equipment is one of the key screening methods for direct estimation.

Solubilization of Hydrocarbons BSs also impact bacterial cell surface hydrophobicity by initiating structural changes in cell surface which ultimately increases the availability of the hydrocarbons to bacterial cells [7, 9]. Bacterial and microbial adhesion to hydrocarbon is indirect evidence of the BS production of any microbial strain.

The Hydrophilic-Lipophilic Balance (HLB Value) A distinguishing feature of BSs is the hydrophilic-lipophilic balance (HLB) which stipulates that the fraction of hydrophilic and hydrophobic moiety in surfactants [7]. The HLB value is a degree to designate whether BSs are associated with water-in-oil (w/o) or oil-in-water (o/w) emulsion (Fig. 2.1) [10]. The surface activity in terms of HLB value makes BSs excellent bio-emulsifiers, foaming, and dispersing molecules [11]. Therefore it can be concluded that BSs effectiveness can be measured by determining its potential to reduce surface and interfacial tension [4].

2.2   Introduction to the Screening Concept

Interest in biosurfactants (BSs) research into and applications is gaining increased thrust because of their environmentally friendly properties and lower toxicity in contrast to chemical surfactants [12, 13]. The search for new BSs producing strain is a key area in surfactant research and applications. The availability of efficient methods for the screening of BSs production and the consequent assessment of

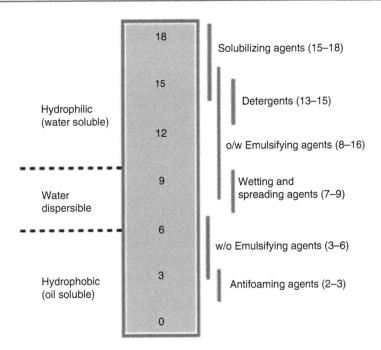

Fig. 2.1 HLB scale representing a different class of surfactants

surface activity holds the key to the discovery of novel BSs. The development of BSs with high value-added characteristics is a vital part of future research development which is strengthened by structured screenings and computational modelization [14]. This chapter presents a summary of the present approaches with their advantages and disadvantages for the isolation and screening of BSs producing microbes. There are various qualitative and quantitative methods with variable precisions and aims for the screening of effective BSs producers.

The inclusive adaptation of BSs is recognized to be obstructed by inadequate information and a lack of accessibility to economically viable BSs. Presently, there is only very inadequate information about commercially existing BSs, e.g. rhamnolipids, surfactin, and sophorolipids. A range of novel BSs producing microorganisms is the key problem in incapacitating the economic hindrances of the production. So, improved efforts in the detection of new BS producing strain must be made by using an extensive range of various screening approaches, which is the prime emphasis of this chapter.

The major aim in screening for novel BSs is identifying structural composition with efficient interfacial potential, low critical micelle concentration (CMC), significant emulsion forming potential, better solubility, and functionality in an extensive pH-range. BSs should be commercially versatile as well as economically inexpensive. Consequently, the secondary level of extensive screening is the detection of efficient producers with maximal yields. Since the 1970s, an extensive range of

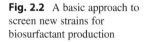

Fig. 2.2 A basic approach to
screen new strains for
biosurfactant production

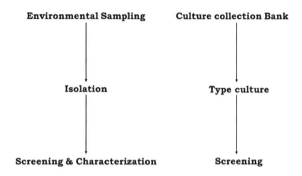

direct and indirect approaches for the screening of BS producing strains has been
established and successfully utilized. All direct and indirect methods have typically
been limited to a practicable number of samples. Due to the increasing demand for
greener BSs, it is of interest to isolate, characterize, and functional assessment of
novel strains producing BSs from environmental sources. In the last decade, surface-
active agents of microbial origin are being explored in various food,
pharmaceuticals, and agricultural applications due to their origin and ecofriendly
nature.

The present inclination of utilizing more natural molecules than the chemical-
based due to their beneficial impact on health and the environment has also shaped
demand for "green" surfactants. A variety of novel BSs producing microorganisms
are the key concern in incapacitating the cost factor hindrances of commercial
production. Hence, increased exertions in the detection of new BSs producing strains
must be made by executing an extensive range of various screening approaches,
which is the key to the present chapter.

A comprehensive application of such tactics could ultimately lead to the
anticipated increase of new commercially exciting strains. An effective strategy to
screen large microbial populations is the vital aspect to accomplish isolating novel
and efficient strains because many strains need to be identified. A broad approach for
screening of new BSs or producing microorganisms can be summarized in three
steps: sampling, isolation strategies, and characterization of strains (Fig. 2.2).

A highly efficient BSs strain can intensify the competitiveness of the investigation groups. High-throughput screening (HTS) methods play a crucial role in
reaching for efficient strains, commercial production, low cost, and high specificity
[15]. Chances for getting the efficient BSs producing microorganisms depends upon
the proper sampling and isolation of appropriate microorganisms are the foundation
for the screening of BSs. According to Ron and Rosenberg [16], BSs are involved in
achieving different physiological roles and deliver many benefits to producing
microorganisms:

- Emulsification of substrates-helps in the uptake of insoluble nutrients
- Uptake of hydrophobic nutrients
- Anionic surfactants are known to play a role in heavy metal bindings

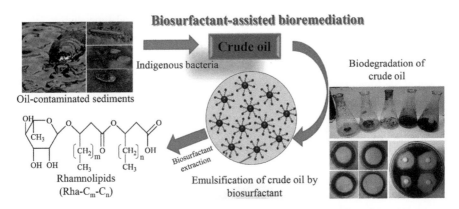

Fig. 2.3 Isolation of BSs producing bacteria with subsequent application in bioremediation Wan [29]

- Possibly involved in pathogenesis mechanisms like biofilm formation
- Provides antibiosis effects against competitive interactions against neighboring microbial populations
- Helps in the colonization of the strain at various natural niche such as the gastrointestinal tract
- Degradation of polycyclic aromatic hydrocarbons

As the BSs producing strains are involved in different roles, isolation of new strains can be obtained from various sites such as hydrocarbon contaminated [17], foodstuff [18–20], spoiled waste material [21], gastrointestinal surfaces [22], fermented foods [23–25], extreme environments like halophilic [26], and marine environments [27]. In divergence, various undisturbed sites have yielded many interesting BSs producers. However, the chances of getting efficient BSs producing stains from affected sites as compared to the uncontaminated sites [28]. Bioremediation using oil-degrading microbes has appeared as a capable green cleanup substitute in recent times. The utilization of BSs producing and hydrocarbon-degrading indigenous strains boost the efficiency of bioremediation due to the bioavailability of hydrocarbons due to surfactants' action (Fig. 2.3). So, isolation of the efficient biosurfactant producer from hydrocarbon-contaminated sites increased the chances to get a high-yielding producer.

In the ensuing steps, the microbial populations must be identified, for further confirmation of surface active properties. Various methods have been documented for identifying BSs producing microorganisms. The majority of methods depend upon the surface and interfacial properties of the amphiphilic molecule. Observation of the surface and interfacial tension reduction is the direct quantitative approach (Fig. 2.2). Direct methods have the advantage of fast screening with accurate measurements.

On the other hand, qualitative screening includes various in-direct methods such as drop collapse, hemolysis, oil displacement, and BATH assay. Such kind of methods are cost-effective, required very limited resources, and time-saving. With recent development and progress in engineering, microfluidics, and robotics in the miniaturization of a process in biological system results in the development of high throughput screening approaches (HTS). The process of HTS for assessment of large potential microbial populations or axenic type culture collections is imitated in the end (Fig. 2.4).

2.3   Isolation of Biosurfactants Producing Microorganisms

The step by step process of new BS discovery starts with the isolation of microbial strain followed by screening, structural characterization, and identification of producing microorganisms (Fig. 2.2). There is an extensive range of traditional, commonly established methods for the isolation and determining of the microbial population for BS producers. Different autochthonous and allochthonous microorganisms (yeast, fungi, bacteria, and actinomycetes) have been isolated from various environmental sites such as contaminated soil, effluent contaminated water bodies, and freshwater, collected from the globe. The chances to get the BS producing environmental isolates rise to the level of 25–80% from contaminated sites using hydrocarbons enrichment. In actual only a small percentage of environmental strains produce BS. Interestingly, this number can increase up to 25% in samples obtained from hydrocarbon or from heavy metals-polluted sites or even more than that of 80% when enrichment cultures containing hydrocarbons are used [30–34].

The isolation method is a critical aspect to get efficient BS producer strain. Furthermore, the cultural conditions and nutrient media explored for the initial isolation and cultivation of the environmental isolates can influence the behavior of microorganisms and hinders BS production. Differences in the nutritional requirement, membrane transport capacity, and enzymes involved in the breakdown of the nutrient are different in every strain. Therefore, it is expected that the distribution of strain capable to produce BS in different cultural conditions and environments is underestimated. To counter the problem of cultural-based isolation of the BS producer from a given environmental sample, the metagenomic methods have been extensively employed for the screening of BS. In recent attempts, screening of lipopeptides production using metagenomic approaches has been established. The genes present on non-ribosomal operons for lipopeptide productions have been traced using gene primers and the possibility of utilizing various databases and bioinformatics tools [35]. Metagenomic kind of approaches is the most inexpensive and time-saving to analyze BS productions in various environments. But it is only limited to the functionally known BS types. Thus, such methods can be useful for determining the incidence of BS producers in diverse ecological niches as well as finding "unknowns" of BS screening. As a substitute, media enrichment is a conventional approach for the successful isolation of the BS producers. The enrichment method is the utilization of various nutrients to favor the growth of a specific microorganism over unwanted strain, enriching a strain of the interest. Enrichment

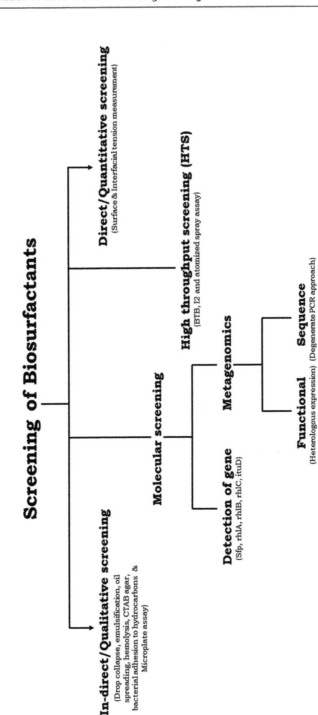

Fig. 2.4 A comparative strategies to screen microbial populations

of the BS isolation medium with hydrocarbons as an enriching nutrient has increased the chances to many folds.

In the culture enrichment method, environmental samples are used to enrich in a liquid medium with hydrocarbon as the sole carbon source to enhance the isolation of degradative strains. Furthermore, the enrichment sample, isolation has been performed using plate agar methods. Pure colonies obtained on agar plates are subjected to the BS screening using different direct and indirect methods (Fig. 2.2).

2.4   Screening of Biosurfactants

The production of BSs with high value-added potential is a major part of research development which is strengthened by using computer-assisted HTS methods and modelization [14]. Various methodologies based upon the physical attributes of the BSs are used to potent BS producer based on various qualitative and/or quantitative methods. On the other hand, the strains with the ability to interfere with hydrophobic interfaces can be discovered. At the same time, some of the extensive screening approaches such as blue agar or CTAB agar are only appropriate to a limited class of BSs. BSs screening protocols are qualitative and quantitative approaches. Primarily, qualitative screening based upon the indirect physical properties is generally enough for screening a large number of microbial populations.

The qualitative screening methods are only helpful for the detection of BS producers. But for the estimation of surface and interfacial tension is a concern of quantitative determination with precision. In the literature, various qualitative methods are explored to screen BS; (1) hemolytic assay [36]; (2) N-Cetyl-N, N, N-trimethylammonium bromide (CTAB) blue agar plate method [37]; (3) bacterial or microbial adherence to hydrocarbons (BATH, MATH) [7, 37]; (4) drop collapse method [38]; (5) oil spreading assay [39]; (6) emulsification assay [40]; (7) microplate assay [41]; (8) penetration assay; (9) colorimetric test (Siegmund and Wagner [37]; (10) solubilization of crystalline anthracene; (11) salt aggregation assay Lindahl et al. [42]; (12 Replica plate assay Mozes and Rouxhet [43] (Table 2.1). Most of the qualitative assays require time to complete in a few minutes to hours or maximum a day for execution.

On the other hand, different quantitative methods such as direct surface and interfacial tension measurement by axisymmetric drop shape analysis (ADSA) [1–3], du Noüy ring method, and Wilhelmy plate method are the gold standard assays in surface chemistry. Some of them are semi-quantitative methods like the emulsification index and emulsification assay Cooper and Goldenberg [40].

In the recent developments, the HTS method such as the atomized drop method [45]; dynamic surface tension measurement [48], and bromothymol blue assay are well demonstrated and reproducible [58]. In recent literature, the various researcher has compiled the screening methods for microbial BSs [1–3]. Screening of the most known glycolipid BSs, i.e. rhamnolipids has been documented by the Varjani et al. [8] (Fig. 2.5). The author has effectively demonstrated how initially the screening of rhamnolipids starts with qualitative, quantitative, and HTS methods. Furthermore,

Table 2.1 Comparison of different methods for biosurfactants screening (Modified and adopted from [32])

Analytical Technique	Qualitative Analysis	Quantitative Analysis	Analysis Speed	Application in HTS	Reference
Direct surface/ interfacial tension measurement	++	+	Min	−	Mnif and Ghribi [44]
Atomized drop method	++	++	Days	+	Burch et al. [45]
Axisymmetric drop shape analysis	++	++	Min	+	Rotenberg et al. [46], van der Vegt et al. [47]
Dynamic surface tension measurement	++	+	Min	+	Bakshi [48]
Drop collapse assay	++	−	Min	+	Jain et al. [38], Bodour and Miller-Maier [28]
Hemolysis assay	+	−	Days	−	Banat et al. [49]
Orcinol assay					Dwivedi et al. [50]
Anthrone assay					Tuleva et al. [51]
Microplate assay	++	−	Min	+	Vaux and Cottingham [41]
Penetration assay	++	−	Min	+	Singh and Sedhuraman [52]
Colorimetric test	+	−	Min	−	Siegmund and Wagner [37]
Oil spreading assay	++	−	Min	−	Morikawa et al. [39]
Emulsification capacity assay	+	−	Days	−	Cooper and Goldenberg [40]
Solubilization of crystalline anthracene	+	−	Days	+/−	Kim et al. [53]
Bacterial adhesion to hydrocarbons assay	+	−	Min	−	Dillon et al. [54]
Hydrophobic interaction chromatography	+	−	H	−	Sharma et al. [55]

(continued)

Table 2.1 (continued)

Analytical Technique	Qualitative Analysis	Quantitative Analysis	Analysis Speed	Application in HTS	Reference
Replica plate assay	+	−	Days	−	Mozes and Rouxhet [43]
Salt aggregation assay	+	−	Min	+/−	Lindahl et al. [42]
CTAB agar assay	+	−	Days	−	Shoeb et al. [56]
Hemolysis assay	+	−	Days	−	Thavasi et al. [57]
Bromothymol blue (BTB) assay	++	++	Min	+	Ong and Wu [58]

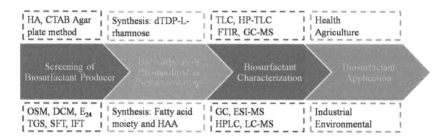

Fig. 2.5 A model approach for rhamnolipids screening (Courtesy by: [15])

for an efficient rhamnolipids production, screening of high titer yielding strain with homogenous rhamnolipids type is a crucial step.

2.5 Qualitative/Indirect Methods

2.5.1 Agar Surface Overlaid with Hydrocarbons

Pure cultures are streaked using an inoculating loop on hydrocarbon coated agar plates and the same plates are incubated for 5–7 days at 37 °C. All the positive colonies encircled by a halo of the hydrocarbon emulsified zone are observed as BS producers [59]. Agar plate overlaid with hydrocarbon is a rapid and efficient screening method where the appearance of the zone of emulsification around the isolates is the direct sign of BS production (Fig. 2.6). The average zone diameter of the emulsified area can be considered as BS producing ability. Budsabun [61] demonstrated that the agar plate overlay method is useful in screening *Serratia marcescens* derived BS.

Fig. 2.6 Agar plate overlaid
with hydrocarbon for BS
screening [60]

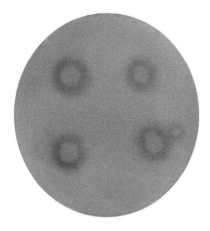

Advantages The process is simple, rapid, and can be used for a large number of isolates at the same time. No sophisticated laboratory facility is required to execute the above-mentioned BS screening.

Limitations Specific BS cannot be detected using the agar overlay method. There is no discrimination in the activity for the BSs or emulsifiers. One cannot able to differentiate the activity of the low molecular weight (LMW) and high molecular weight (HMW) BSs.

2.5.2 Blue Agar Plate for Extracellular Glycolipids

It is a semi-quantitative BS screening method with a special application in the screening of anionic glycolipids to isolate efficient strain. The blue agar plate allows users to identify BSs producing isolates colonies by the color reaction. Blue agar plate is commonly used for the quantitative screening of extracellular rhamnolipids. In the present assay, BSs develop blue color in minimal salt medium with forming insoluble complex with the cationic surfactant, i.e. cetyltrimethylammonium bromide (CTAB), and the dye methylene blue which was included in mineral salt agar (MSA) plates. Rhamnolipids producing *Pseudomonas aeruginosa* develops dark blue halos around the colonies (Fig. 2.7). Blue agar plate test can be used for all the other anionic glycolipids for BS screening.

Briefly, MSA supplemented with sugar supplemented with glucose (2%) and CTAB: 0.0005% and methylene blue (MB: 0.0002%) [37]. The isolated strain has been cultivated on a blue agar plate containing and if anionic BSs are produced, they form dark blue halos.

Chemical anionic surfactant can be determined by the formation of an insoluble complex with different cationic substances, such as inorganic metal salts, basic dyes (safranin), and cationic surfactants. Salts like $CaCl_2$ and $Al(OH)_3$ form precipitation zone around the colonies in the presence of rhamnolipids.

Fig. 2.7 Blue agar/CTAB
plate screening assay for
glycolipid BSs

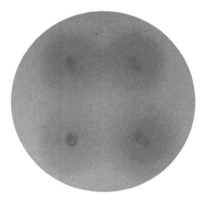

Advantages The blue agar or CTAB-methylene plate assay is a specific method for the detection of anionic glycolipids. Its advantageous to screen large microbial isolates of any environmental samples to detect the anionic surfactants and one can apply various cultural conditions like supplementation of the desired carbon source. The color change is because of insoluble ion complex rather than pH change or production of organic acids by the isolates. Similar kinds of observations are observed with the other anionic BSs such as sophorolipids. So the CTAB agar screening could be applicable for the detection of other glycolipids obtained from the other microorganisms.

Limitations CTAB is antimicrobial or inhibitory in nature for various bacterial strains, so a modified medium with other cationic surfactants can be used for the screening of the BSs.

2.5.3   Hemolytic Activity

The hemolytic activity of BSs was first observed for the surfactin produced by the *Bacillus subtilis* leads to the red blood cell lysis [62]. Biosurfactants are known for their detergency properties and behavior. Therefore, hemolysis has been explored as a primary selection criterion for the screening of surfactant-producing microorganisms. Hemolysis is a type of qualitative screening for the detection of BS production. Approximately, 5% (v/v/) sheep blood supplemented to the nutrient agar medium before solidification of agar plates [36]. Isolates can be spotted or single line streaked blood agar plates and further incubated for 48 h. Blood hemolysis with clear red blood cell (RBC) lysis can be observed for BS screening (Fig. 2.8). Hemolysis is the result of the cell membrane rupture due to the presence of detergency showing molecules.

Hemolysis of RBC has been used to screen and quantify rhamnolipids and surfactin blood agar lysis has been utilized to screen surfactin and rhamnolipids [63, 64]. There is a relation between hemolytic activity and the presence of

Fig. 2.8 Hemolysis of RBC
on the blood agar plate

surfactant, and it has been recommended to use the blood agar lysis as a key assay to screen for BS activity Carrillo et al. [36].

Advantages The blood agar method or hemolysis due to the production of BS is a preliminary screening assay. It is a qualitative indirect method, which is cost-effective, time-saving and can be used to rule out non-BS producers from a huge microbial population.

Limitations The blood agar method is having various limitations. First, the screening assay is not precisely not only lysis blood cells, as various lytic enzymes can also form clearing zones. Furthermore, hydrophobic substrates such as medium enrichment with hydrocarbons cannot be involved as a sole carbon substrate. Also, the diffusion behavior of the BSs can impede the development of clearing zones. Besides, Schulz et al. [65] demonstrated that that various BSs do not show any hemolytic behavior at all, probably due to the low detergency capacity of some of the BSs. Though, its also evident that only 13.5% of the hemolytic isolates reduce the surface tension below 40 mN/m.

2.5.4 Drop Collapse Method

Cell-free supernatant (CFS) containing BSs will be incapable to form stable drops on an oil-coated surface. Whereas, CFS with no surface active properties holds the drop shape on an oil-coated surface. The drop collapse assay is based on the capacity of BSs to disrupt liquid droplets on an oil-coated surface, and this potential is interrelated with its surface tension. Drop collapse assay depends upon the destabilization of liquid suspension droplets by BSs. If the CFS contains the BSs, the polar water molecules are repelled from the hydrophobic surface and the water drops remain stable (Fig. 2.9). And if, BSs s present, due to which the surface and interfacial tension of the drop on oil-coated surface is decreased. Surfactants' concentration and surface tensions are responsible for the stability of the drop. The drop-collapsing

Qualitative – Pennzoil®

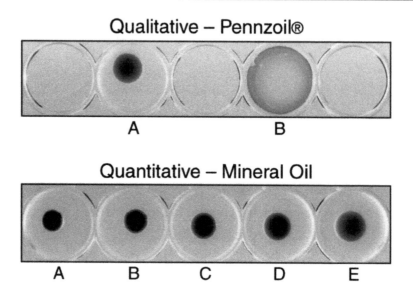

Quantitative – Mineral Oil

Fig. 2.9 Drop collapse assay (Qualitative): (**a**) Water, (**b**) Rhamnolipids 1000 mg/L. Quantitative drop collapse assay (**a**) Water, (**b**) Rhamnolipids 25 mg/L, (**c**) 50 mg/L, (**d**) 75 mg/L, and (**e**) 100 mg/L rhamnolipid (Courtesy by: [28])

method is much sensitive than the blood hemolysis [38] for screening BSs producing isolates. No hemolysis zone has been reported after 24 h of cell growth on the blood agar plate, whereas drop collapse has been observed for all the isolates.

It was well reported that the use of a glass surface as a replacement for the oil surface gives a better improvement in the observations [66]. Even the drop collapse method can be used to quantify the BSs concentration in a given sample by recording Bodour and Miller-Maier [28].

Advantages Cell-free suspensions of the isolates caused the collapse of drops on oily surfaces, demonstrating that even low quantities of BSs, and can be detected by this method. A significant distinction of this method is that it can be upgraded to an automated screening method using microplates. The drop collapse essay is fast and easy to be carried out, requires small volumes of samples with no specialized equipment.

Limitations The drop collapse assay is not appropriate for screening colonies having high emulsifying potential and which do not reduce surface tension considerably. But it displays a relatively low sensitivity because a large number of surface-active compounds should be present to cause the aqueous drop to collapse on the oily or glass surfaces. However, the drop-collapsing method is beneficial for the rapid screening of strains that produce surfactants.

2.5.5 Oil Spreading Assay

Oil spreading assay is one of the easiest and rapid methods to distinguish the presence of BSs in CFS. If the BSs are present in the CFS, the motor oil is displaced and a clearing zone is developed. The measurement of the diameter of the clearing zone can be related to surfactant potency, which is known as an oil displacement activity. Briefly, a small quantity of crude motor oil is added to 50 mL of distilled water in a Petri plate. 10μL of CFS is transferred on the oil-coated water surface (Fig. 2.10). A colony-forming an emulsified halo is regarded as a positive strain for BS production [39].

Advantages Present assay does not require specialized equipment and only requires very little volume. It can be used when the surface activity and BSs quantity are low and can be utilized to screen a large number of samples [67].

Limitations The potential to detect BSs production in different strains has not been checked.

2.5.6 Emulsification Index (EI)

Emulsification is a process of the formation of the liquid, considered as an emulsion, which contains tiny non-coalescence droplets of oil [68]. The formation of the emulsion can be measured by calculating EI [40]. EI is measured by calculating the emulsion height. Briefly, kerosene oil is added to CFS (1:2 v/v), mixed vigorously for 2 min, and permitted to stand for 24 h (Fig. 2.11). The height of the emulsion developed is measured by recording the layer develops in between the aqueous and kerosene layer. Various reports about the utilization of the EI method to select BSs producer strains have been documented [69, 70]. Higher is the surfactant concentration, the higher will be the EI stability.

Fig. 2.10 Oil displacement assay

Fig. 2.11 Emulsification
assay

$$E_{24} = \text{Height of the emulsion layer/total height} \times 100$$

EI method can be applied in different screenings, where the kerosene can be substituted with other hydrophobic substances, e.g. hexadecane. But the correlation between surface activity and emulsification capacity does not occur always.

Advantages Evaluation of emulsification capacity is an effective approach to screen microbial strains. It is a time-efficient and cost-effective approach.

Limitations All the emulsifiers producing strain can be detected using the emulsification method, but all the surface-active agents, i.e. biosurfactants are poor in EI. Therefore, this approach gives just an indication of the presence of BSs. So, one method to distinguish and current research in the direction of identification should rely on a broad spectrum of assays and not only on surface tension observations, which are frequently used as a primary assay.

2.5.7   Solubilization of Crystalline Anthracene

Solubilization of polycyclic aromatic hydrocarbon (PAH) can be utilized to screen the microbial population of BSs producers. Solubilization of the PAH can be sued to select BSs producers [71]. Crystalline hydrocarbon, anthracene solubilization can be used to screen BSs producers. So, crystalline anthracense is supplemented to the CFS and incubated at room temperature for 24 h. This can be quantified on the estimation of the anthracene using spectrophotometric absorption at 354 nm and relates to the production of BSs.

Advantages Solubilization of the PAH is a simple and rapid screening approach, and can be used to screen large populations.

Limitations In certain cases, the anthracene might be toxic to microbial growth.

2.5.8 Turbidity Assay

Turbidity assay was used by Rosenberg et al. [72], which was further modified by Neu et al. [73]. CFS is filtered through membrane filters and the supernatant is freeze-dried and suspended to a buffer and absorption was measured at A_{446} nm. Hexadecane is added to the suspension and permits to stand for the next 10 min at room temperature and absorption was again measured. Later, the accuracy of the turbidity assay was improved by estimating unknown BSs [74].

Advantages A UV/VIS spectrophotometer is the only technical equipment requires in the turbidity assay. This simple approach is effective in time-saving, accurate, and cost-effective quantification of crude BSs obtained from diverse bacteria.

Limitations It is a time taking process for the evaporation and drying of the samples, which makes this approach inappropriate for the rapid estimation of BSs.

2.5.9 Microplate Assay

The surface activity of any isolate can be resolved qualitatively with the use of microplate assay [41]. Microplate assay depends on the modification in optical alteration that is instigated by BSs in an aqueous solution. In the presence of water, a hydrophobic well has no optical delusions or flat surface. The incidence of BSs surfactants reasons some wetting at the corners of the well and the liquid surface converts to concave and attains the shape of a deviating lens.

Briefly, a small volume of the CFS of the testing isolate is taken and put into a 96 microwell plate. Furthermore, the microwell of the 96 well plates is observed using a backing grid sheet. If BSs are present, the distorted images turn out like a concave surface (Fig. 2.12). It is a qualitative assay for the detection of the BSs. The

Fig. 2.12 Microplate assay for the screening of BSs

present method is easy, rapid, and sensitive and permits prompt detection of BSs. Moreover, the assay is appropriate for automated HTS screening. Chen et al. [75] established the efficiency of the microplate assay for high-throughput screening resolutions. The effectiveness of the approach was established by the screening of a bacterium producing BSs and the potential to reduce the surface tension.

Advantages The method is found to be very sensitive, fast, and easy to accomplish than other known methods. It does not prerequisite specialized equipment or substances and eliminates the prejudice which obtains from the BSs properties.

Limitations There is no quantitative data available for BSs screened.

2.5.10 Cell Surface Hydrophobicity (CSH)

Rapid characterization of BS producing bacterial isolates was attained by assaying CSH which had a direct relationship with BS production. CSH can be analyzed by different mechanisms, such as hydrophobic interaction chromatography (HIC), salt aggregation test (SAT), bacterial adherence to hydrocarbon (BATH), and adhesion to polystyrene surface using a replica plate test, gives a simple way means for identifying bacteria for the production of BSs.

2.5.11 Measurement of CSH

It is a direct correlation between cell surface hydrophobicity (CSH) and BS production. CSH plays a vital part in the attachment and detachment from the surfaces. High CSH allows the strain to attach to hydrocarbon droplets on the surface of cells and to move from water to organic, hydrocarbon phase, where BSs and enzymes decompose hydrophobic wastes [76]. Briefly, the microbial cells are cultivated and harvested by centrifugation and resuspended in the buffer to absorption(A_{600}) value of approximately 0.5. Subsequently, cell suspensions are mixed to hydrocarbons and mixed for 5 min and permitted to settle for 15 min for the hydrocarbon phase. Furthermore, the aqueous phase is used to measure adsorption(A_{600}) [77]. The reduction in absorption of the aqueous phase is considered as a degree of the CSH (H%), which is determined as follows:

$$H\% = [(A_0 - A)]/A_0 \times 100$$

where A_0 and A were A_{600} before and after mixing with hydrocarbon, respectively.

Advantages CSH determination is a simple assay that suggests the potential to adhere to hydrocarbon can be considered as a characteristic behavior of BS producing microorganisms.

Limitations   In certain cases, the ability to adhere to hydrocarbons by the isolates is not limited to hydrocarbon-degrading microorganisms but can also be observed in the example of *Serratia marcescens* which is incapable to degrade hydrocarbons. Production of short-chain fatty acids by bacteria can also lead to mistaken results.

2.5.12   Hydrophobic Interaction Chromatography (HIC)

HIC is a chromatographic assay based on hydrophobic interaction among nonpolar functional groups on a gel bed and nonpolar regions of a solute. Such studies with hydrophobic gels advised the possibility that HIC might give a simple and beneficial method to differentiate between microbial strains about the hydrophobicity of their surfaces or adhesins [78]. Briefly, the phenyl sepharose CL-4B, bed volume about 0.6 ml, as the column packing matrix. Furthermore, the column was saturated with a solution of NaCl in citrate buffer. The cell fraction was prepared which assisted in equilibrating the gel. Subsequently, a small sample was introduced on the gel trailed by an equilibrating solution. The non-retained cell was compared with the original cell suspension by determining the absorbance at A_{540} nm and the results were expressed as the percentage of the retained cell.

2.5.13   Salt Aggregation Test (SAT)

SAT is considered a highly specific method to determine the CSH in highly pathogenic microorganisms [42]. Briefly, a cell suspension was mixed with an equal amount of salt suspension. The cell/salt suspension was moderately mixed at room temperature and visual observations should be recorded compared to a black background. The findings are articulated as the lowest concentration of salt causing cell aggregation.

Advantages   The SAT analysis screening method offers a simple, rapid, and reproducible protocol for determining CSH well suitable for screening of large populations in a definite time frame.

Disadvantages   Interestingly, in some cases, sal aggregation efficiency has been decreased or lost on subculturing for up to 20 times in the case of *Streptococci* [79].

2.5.14   Bacterial Adherence to Hydrocarbons (BATH)

BATH is a modest and rapid method for measuring CSH. BATH assay has found utilization in the determination of the surface properties of extensive diversity of bacteria and bacterial populations. Connections have been found among the adherence of bacteria to hydrophobic substances and their attachment to other surfaces, such as polystyrene surfaces, epithelial cells, and teeth surfaces. Briefly, cell suspensions prepared at regular intervals in phosphate buffer were distributed in test tubes. Hydrocarbons such as hexadecane were then added and incubated at room temperature for 20 min to let hydrocarbon separation and the absorption of the aqueous phase was determined initially and final after treatment. All the observations were logged as the percentage of absorbance of the aqueous phase at initial and final values [80].

Advantages The ease of measuring CSH delivers a rapid assay for determining BS formation before their isolation and further assessment. The benefit of this assay lies in its low-cost way to detect a display of bacterial strains for BS production concurrently on readily available materials.

Limitations The imitation of the BATH method is lies when bacterial culture exposed to diverse cultural conditions such as nutrient composition, air transfer, pH, age of culture, and temperature.

2.6   Measurement of Surface Tension

The direct measurement of the surface and interfacial tension of the cell-free supernatant is the most upfront screening approach and very suitable for the initial screening of BSs producing strain. There are different standard approaches to determine surface and interfacial tensions. Direct surface tension measurement using the Du-Nouy ring method, Wilhelmy plate method, Stalagmometer, Pendant drop method, and Axisymmetric drop shape analysis (ADSA). All these methods can be utilized for determining the surface and interfacial tension of a cell-free supernatant.

2.6.1   Tensiometeric Measurement of SFT

The tensiometer is an equipment used for the determination of the surface and interfacial tension of a cell-free extract. The determination of surface tension using a tensiometer instrument is one of the appropriate methods. Determination of the surface tension is not possible to use for a large population at the preliminary screening level.

2.6.2   Du-Nouy Ring Approach

The Du-Nouy ring approach is mainly based on the determination of the force essential to detach a platinum ring from a surface of the interface. The required force to detach the ring is directly proportional to the surface and interfacial tension. It involves gradually lifting a ring, generally made up of platinum, from the surface

Fig. 2.13 Schematic diagram
of the Du-Nouy ring method
(Courtesy by: https://www.
kruss-scientific.com)

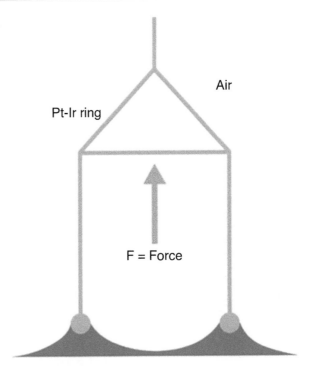

of a cell-free supernatant in case BSs (Fig. 2.13). The Du-Nouy ring method was demonstrated by the French physicist Pierre Lecomte du Noüy in 1925.

The platinum ring is used as the reference material for the preparation of the ring, as it forms a contact angle θ of $0°$ with liquids. Surface and interfacial tension can be determined using an automated surface tensiometer which is available in different specifications and variants. The Du-Nouy should be free from any contaminant, sterilization of the ring can be attained using a platinum ring. The Du-Nouy ring method is extensively utilized for the screening of BS producing strains. It is recognized as an isolate that is promising if it decreases the surface tension of a suspension to 40 mN/m or less. It is uniformly assumed that an efficient BS producer is considered as one being able to decrease the surface tension of the suspension by ≥ 20 mN/m as compared to the distilled water. The benefit of the ring method is the precision and ease of handling. Though, it needs specific physical tools and equipments particularly with high precision.

Advantages Du-Nouy ring method is used to measure the surface and interfacial tension of the cell-free extract. In particular, the Du-Nouy ring approach is relatively easy and most often utilized.

Limitations A common drawback is the ring measurements of surface tension of different samples cannot be executed simultaneously. Subsequently, the minimum volume to be utilized for the measurement of the surface tension using the ring method is high. It can only measure the surface tension with an appropriate volume of the sample. The Du-Nouy ring method can only measure the surface tension in a quasi-static equilibrium. Furthermore, high-viscosity suspension leads to high stress on the fragile platinum ring to a certain degree.

2.6.3   Wilhelmy Plate Method

It is important to determine surface and interfacial tension changes over time, (*measurement at different times, and end-point SFT and IFT determination for CMC concentration*). *Here* Wilhelmy plate can be a choice of strategy, as compared to the Du-Nouy ring method. At the same time, high-viscosity suspensions cause higher stress on the fragile Du-Nouy ring method. The Wilhelmy technique is favored for such type of applications. Both approaches have a major difference that the Du-Nouy ring is dragged through the surface to observe the SFT, whereas the Wilhelmy plate method is working on stationary measurements. The force working on a vertically immersed plate is observed and determined. When a vertically immersed plate touches a suspension surface or interface, at that time a force F, which relates with the SFT or IFT σ and with the contact angle θ as per the subsequent equation, which acts on Wilhelmy plate (Fig. 2.14):

$$\sigma = F/L \times \cos \theta$$

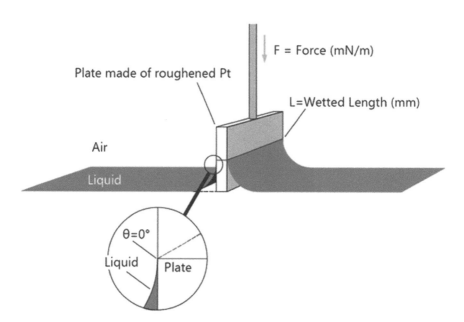

Fig. 2.14 Schematic diagram of the Wilhelmy plate method

2.6.4 Stalagmometric Method

The stalagmometric approach is one of the extensively adopted low-cost approaches for measuring SFT and IFT. The idea is to determine the weight of the suspension drops falling from a capillary glass tube, and in that way calculate the SFT of the sample. One can determine SFT by counting the weight and number of the falling drops. The weight of the drop depends upon the SFT of the suspension (σ), the radius of the capillary glass tube (r), and the acceleration because of the gravitational force as per Tate's law. The SFT of fluid can alternatively be determined using a Traube stalagmometer. The Traube device is a pipette with a wide flattened tip, which allows large drops of the same size to form a reproducible size due to the gravitational force. The SFT can be measured based on the number of drops that fall per volume, the gravity of the fluid, and the SFT of a standard liquid such as water.

Advantages Simple, rapid, and low-cost non-specialized types of equipment are required.

Limitations The limitation of the stalagmometer approach is that only repeated measurements can be performed and seems to be variability prone. This approach can lead to a large inconsistency. Other limitations are the flow rate of the fluid, and the contact period of the fluid or surface age is very limited.

2.6.5 Pendant Drop

The pendant drop or ADSA method is an optical approach for determining the SFT and IFT. A drop of interesting fluid can hang from the capillary end. The drop espouses an equilibrium shape that is exclusively by the tube radius, the SFT and IFT, its density, and the gravity field. The fluid sample can be suspended from a capillary needle in a bulk liquid or gaseous phase. The gravitational force and relationship between SFT and IFT have determined the shape of the drop. The SFT and IFT are measured from the shadow image of a pendant drop employing drop shape analysis. The Young-Laplace fit can also be explored to determine the distortion of a drop on a solid surface for calculating the contact angle in the case of a sessile drop approach.

Advantages Simple, consistent, and free from human intrusion.

Disadvantages The drawback of the pendant drop shape method is again that measurements cannot be achieved concurrently, and interference of the temperature and environmental pressure.

2.6.6   Axisymmetric Drop Shape Analysis (ADSA)

ADSA is an important method for the determination of SFT and IFT along with the contact angles of the pendant, sessile drops, and bubbles. ADSA depends on the Laplacian curves and an experimental profile (Figs. 2.15 and 2.16).

The drop shape technique straightforwardly enables the analysis of both liquid–vapor and liquid–liquid SFT and IFT measurements. The ADSA method is also

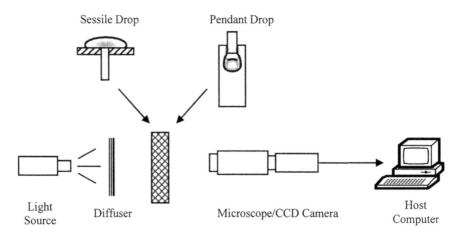

Fig. 2.15 Schematic diagram of the experimental setup of ADSA (Courtesy by: [81])

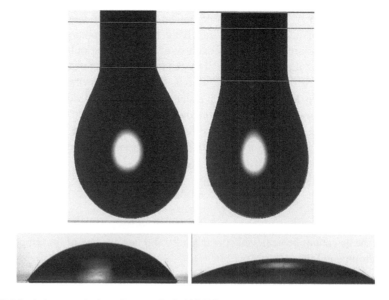

Fig. 2.16 Axisymmetric drop shape analysis (ADSA)

applied to materials that vary from organic liquids to molten metals and from pure solvents to concentrated fluids.

Advantages The ADSA technique is having various advantages as compared to the ring method and Wilhelmy approach as; sample requirement is less as compared to existing or available technique, in combination with the suitable design and automation measures, ADSA can act as a film balance by recording SFT as a function of the surface area of a drop, which is used by changing the drop volume.

Limitations For the ADSA technique, a special setup of a camera and system software is required. The measurement of SFT is rather complex.

2.7 Molecular Tools to Identify Biosurfactant Producing Genes

Exploring the conventional method for the screening of BSs using hemolysis, drop collapse, direct measurement of the SFT, and IFT are routinely used in screening for isolates. Most of the conventional BSs screening approaches are time-consuming, non-specific qualitative approaches, which are laborious, and time taking, and particularly, the identity of the producing strain is also not recognized by them. So quick screening and identification of the newer strains for BSs production by molecular tools is thus a crucial step. Molecular tools for the identification and subsequent screening of BSs producing isolates is an effective method [2]. Various genes associated with the production of the rhamnolipids, surfactin, fengycin, sophorolipids, and many other BSs/BE are known and can be comprehensively determined (Table 2.2). Direct screening for BSs producing genes is quicker and less laborious. A large microbial population from the desired sample can be screened within a time frame with the specific genes involved. Direct BSs producing genes would validate the conservative screening approaches [91]. BSs producing *B. amyloliquefaciens* and *B. circulans* have been used to detect the presence of *sfp* gene locus [92].

The incidence of the *sfp* gene locus identified various strains for BSs production, namely *B. tequilensis*, *B. stratosphericus*, *B. subtilis*, *B. safensis*, *B. cereus*, and *B. firmus*.

The molecular detection of DNA extracts from marine origin Bacillus strains for the incidence of functional *sfp* genes permits the user to recognize, in specific, Bacillus marine isolates with a capable ability for BS production [82]. The *sfp* gene locus in Bacillus strains encrypts phosphopantetheinyl transferase enzyme, which is essential for the production of non-ribosomal secretion of surfactin. The gene *sfp* is recognized as an important gene present on the *srfA* operon which codes for a non-ribosomal peptide synthetase complex also considered as surfactin synthetase.

Table 2.2 Different genes with primers details used for the BSs production

Gene	Primer	Sequence	Reference
Surfactin (sfp)	Sfp-f	5'-ATGAAGATTTACGGAATTTA-3'	Porob et al. [82]
	Sfp-r	5'- TTATAAAAGCTCTTCGTACG-3'	
Surfactin (srfC)	Sur-3f	5'- ACAGTATGGAGGCATGGTC-3'	Kavitha et al.
	Sur-3r	5'- TTCCGCCACTTTTTCAGTTT-3'	[83]
Surfactin synthetase (SrfAA)	Srf-Af	5'- TCGGGACAGGAAGACATCAT-3'	Almoneafy
	Srf-Ar	5'- CCACTCAAACGGATAATCCTGA-3'	et al. [84]
Iturin A synthetase C (ItuC)	ituCf	5'- GGCTGCTGCAGATGCTTTAT-3'	Mora et al. [85]
	ituCr	5'- TCGCAGATAATCGCAGTGAG-3'	
Iturin A (ItuD)	ItuD1f	5'- GATGCGATCTCCTTGGATGT-3'	Gond et al. [86]
	ItuD1r	5'- ATCGTCATGTGCTGCTTGAG-3'	
Bacillomycin (BamC)	Bacc1f	5'- GAAGGACACGGAGAGAGTC-3'	Abdallah et al.
	Bacc1r	5'- CGCTGATGACTGTTCATGCT-3'	[87]
Bacillomycin L synthetase B (bmyB)	bmyBf	5'- GAATCCCGTTGTTCTCCAAA-3'	Frikha-
	bmyBr	5'- GCGGGTATTGAATGCTTGTT-3'	Gargouri et al. [88]
Fengycin (fenD)	fenDf	5'- GGCCCGTTCTCTAAATCCAT-3'	Khedher et al.
	fenDr	5'- GTCATGCTGACGAGAGCAAA-3'	[89]
Rhamnosyl transferase gene	rhlA-F	5'-GATCGAGCTGGACGACAAGTC-3'	Schmidberger et al. [90]
	rhlA-R	5'-GCTGATGGTTGCTGGCTTTC-3'	
	rhlB-F	5'-GCCCACGACCAGTTCGAC-3'	
	rhlB-R	5'-CATCCCCCTCCCTATGAC-3'	
	rhlC-F	5'-ATCCATCTCGACGGACTGAC-3'	
	rhlC-R	5'- GTCCACGTGGTCGATGAAC-3'	

Advantages Molecular or PCR based detection of the BSs producing genes is a reliable, specific, and effective method. Screening of the large population from an environmental sample using PCR based method is a time-saving approach.

Limitations For PCR or molecular-based methods, sophisticated laboratory infrastructure with high-end analysis is required. There is a constant need for highly trained personnel for the execution and analysis of the data obtained. Molecular identification of the desired genes is not available for all the BSs producers, such as genes associated with the new strains of lactic acid bacteria that are not known. So, the gene involved, identification of the pathways, and primers to detect the incidence of such genes is a prerequisite for PCR based methods.

2.8 High-Throughput Screening (HTS)

The use of rapid and consistent assays for screening and selection of BSs producing isolates from a large microbial population of possibly active organisms embraces the key to the detection of new BSs or producing strains. A novel BSs screening method to rapidly screen large numbers of isolates for BSs production using together qualitative and quantitative methods with precision is known as HTS. According to Chen et al. [75], the HTS screening approach for the isolation of BSs producing isolates must accomplish some requirements:

- The capability to identify potential isolates
- The potential to measure quantitatively how effective the BS is
- The potential to screen a large population quickly

The detection of novel BSs and the investigation of the genomics of BSs production would greatly advantage from a quantitative and HTS method. Multiple isolates can be concurrently screened within some seconds, thus permitting thousands of isolates to be screened for BSs production in practically real-time.

High-throughput screening technologies play a key role in reaching for improved strains, which are large scale, low cost, and high specificity [93, 94]. High-throughput screening technologies consist of generating a mutant library in high-throughput miniaturized cultivation platforms, such as microplates [95], shake flasks, micro-bioreactors [96], and using a high-throughput detection platform [97, 98].

Advantages Multiple screening and assessment are effective for a large population with precise output.

Limitations Some of the HTS methods have shortcomings such as poor specificity, demands on specialized equipment, and a large number of samples for analysis. Samples cannot be measured in parallel restricts the application to the high-throughput screening on a large scale.

2.8.1 Atomized Oil Method

A novel BSs detection method was established by using *P. syringae*, which synthesizing the lipopeptide surface-active agents syringafactin. The HTS method has been developed where it was hypothesized that the misting of oil droplets on top of agar plates, for the presence of BSs would reflect the interaction of the oil. Consequently, the bacterial strain of *P. syringae* had grown and develops a light-diffractive halo that was seen near the colonies. In the case of, surfactants production by *P. syringae* the droplets implicit an added uniform, energetically advantageous, hemispherical shape. The oil droplets, which were in direct contact with the BSs, were raised on the agar plate possibly due to the emulsification of the oil and observed more spherical as compared to the agar plate surface away from BSs producing colonies. The whole method has been validated and verified for the various structurally diverse commercially available BSs. Most of the commercial BSs acted similarly, leads to the raised oil droplets, hemispherical in shapes that seemed bright when illuminated.

Captivatingly, when all the BSs were categorized by their HLB values, it was observed that BSs with a low range of HLB values, resulted in bright halos, while those with a high range of HLB values lead to dark halos. In the atomized oil approach, the sensitivity of the method has been validated using a varying concentration of the known commercial BSs to detect the lower concentration of that BSs that was still detectable using this method. In conclusion, the atomized oil method detected BSs at ten fold lower concentrations detected using the drop collapse method. A log-linear connection between the quantity of BSs used on plates and the size of the halo was observed. Thus, a quantitative evaluation of the relative variance in the amounts of BSs in two different prepared samples can be easily predictable.

Advantages Highly selective, quantifiable, and reproducible.

Disadvantages Its lowest detection range is 250 mg/L, making it problematic to screen high-yielding mutants obtained from the low-yield parent strain.

2.8.2 Detection of Lipopeptides Using Bromothymol Blue

The commonly used approaches to screen lipopeptide-containing samples consist of SFT measurement by tensiometry measurement, emulsification, and hemolysis assay [32]. Such methods, though, have restrictions including scalability of the protocols, need for sophisticated equipment, utilization of toxic chemicals, and a high frequency of false-positive or negative analysis [45, 99]. In the present report, bromothymol blue (BTB) is recognized to detect lipids using thin-layer chromatography (TLC) [100], as a fast and HTS assay for recognition of lipopeptides. The surfactin, iturin, and fengycin can be discriminated against in the TLC method and they react with BTB displaying different colors at 1 g/L or below concentrations. Interestingly, the occurrence of lipopeptides can be analyzed directly from the cell-free medium, which also abolishing the purification process hence saving time.

BTB solution was prepared in phosphate-buffered saline solution (pH 7.2). The screening was completed in 96 well Elisa plates with an equal amount of BTB and BSs. The quantitative activity was determined using an Elisa plate reader and absorption was recorded at wavelengths of A_{616} and A_{410} nm. The colorimetric response (CR) (%) was measured as reported earlier [101]. Control of the assay was arranged by mixing BTB with culture media. All the samples were adjusted to pH in the range of 7–8 before the tests as the color change is sensitive to the pH change. The change in color showing a linear quantitative response to all the three different lipopeptides viz. surfactin, iturin, and fengycin) at the concentration of the 0–1 g/L. This method was used on the cell-free extract and partially purified BSs producing isolates and the concentration was verified by HPLC estimation. A simple method has been developed for quantitatively detecting lipopeptides and screening lipopeptide-producing strains.

Advantages BTB estimation of lipopeptides has the advantages of quantitative measurements, easy execution, simple procedure, utilization of non-toxic chemicals and sophisticated equipment, and applicability to a surfactin, iturin, and fengycin. Furthermore, the BTB method presented here does not need an incubation time as compared to the other colorimetric method used for the detection of the surfactin.

Limitations The false-positive observation for the BTB assay might be due to BTB interaction electrostatically with the lipid, phospholipid, and proteins moieties. Such types of problems are due to the interaction of other cellular moieties containing lipids with BTB leads to false-positive results.

2.8.3 Polydiacetylene (PDA)-Based Screening for Surfactin

PDA is a category of conjugated polymers that are used as biosensors for the detection of BSs owing to their exclusive color change possibility from blue to red shade [102]. PDA sensor displays blue to red form is generally estimated in terms of colorimetric response. One can use the CR % semi-quantitatively to estimate the product concentration, the higher the CR % value, the higher will be the surfactin concentration [103]. The development of PDA-based HTS screening method for surfactin relied on the blue to the red color change. Zhu et al. [103] used PDA-based screening of thousands of surfactin mutants and validated the assay (Fig. 2.17). PDA-based method enables the inexpensive production of *Bacillus subtilis* surfactin for its extensive applications.

Advantages Using this method, one can develop a cost-effective approach for screening of the capable high-yielding *Bacillus subtilis* in production media. The effectiveness and ease of the detection of surfactin using PDA vesicles can be used for the screening of the BSs such as glycolipids, lipopeptides, phospholipids, and other surfactants.

Limitations The effects of fermentation on cultural conditions such as variations in pH during fermentation, temperature, and other components that are present in broth have never been determined.

2.8.4 A Rational HTS for the Screening of Sophorolipids

A rational HTS approach can expressively improve the effectiveness of strain screening with high performance. In one rational HTS method, a simple and rapid reaction of the sophorolipids (unsaturated fatty acids) and I_2 molecules was developed which established a correlation coefficient (R^2) of 0.9106 with HPLC analysis.

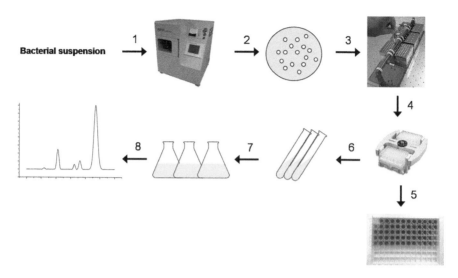

Bacterial suspension

Fig. 2.17 Approach for screening surfactin using PDA vesicles

Furthermore, the addition of chlorpromazine also established a rational selection pressure of synthesis of the sophorolipids due to the key enzyme presence and expression of the P_{450} enzyme also increased the effectiveness. So, rational selective pressure in HTS enabled users to efficaciously screened out high-yielding sophorolipids producers from 1500 isolated colonies, which displayed improvements of 40.3% and 11.4% on concentration and yield, as compared to the wild strain in a fermentor (Fig. 2.18).

Briefly, 20μL of the *Candida bombicola* biomass suspension was treated by atmospheric and room temperature plasma (ARTP) for 1 min with the lethality of about 90%. Furthermore, the cell suspension was properly spread over the agar medium surface containing chlorpromazine. Mutant isolates with the marginally larger size were further selected and introduced into 96-well plates for 24–48 h to get the seed pre-culture for the bioreactor. Cell-free supernatant was obtained intermittently for the estimation of the sophorolipids by the I_2 method.

Advantages The conclusion of rational HTS screening is supposed to be readily demonstrated to the screening of other BSs. The rational HTS approach for the estimation and quantification of sophorolipids using I_2 as the reagent has been observed to be an effective, simple, and rapid method.

Limitations Estimation of the lipid moiety is having a chance of human error and precision also varies from person to person. Determination and identification of the rate-limiting enzymes' feedback inhibition and their regulation are not feasible for all the other known BSs. Identification of the key enzymes is required for the

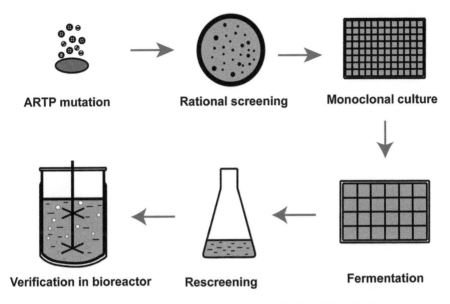

ARTP mutation   **Rational screening**   **Monoclonal culture**

Verification in bioreactor   **Rescreening**   **Fermentation**

Fig. 2.18 Flowchart of HTS rational screening of sophorolipids from *C. bombicola*

successful estimation of the BSs needs to be determined for the reproducibility of the efficacy of this method.

2.8.5 Determination of Carbon Source Concentrations by Cu(OH)2 Method

The mutant of the *Candida bombicola* has been prepared using ARTP method. The I_2 and $Cu(OH)_2$ protocol was utilized for screening sophorolipids and carbon concentrations, respectively. Besides, 2-deoxy-d-glucose was utilized as the rational screening compression, which permitted us to screen high sophorolipids with co-metabolization of carbon source, i.e. glucose and glycerol. Under alkaline pH, polyhydroxyl compounds form a complex of navy blue color on the reaction with Cu $(OH)_2$ solution at A_{630} nm (Fig. 2.19). Therefore, glucose and glycerol in the production medium will form a complex with $Cu(OH)_2$ solution and such an approach could be used semi-quantitatively and qualitatively to determine the utilization of nutrient sources [105].

Advantages A rational screening of sophorolipids using 2-deoxy-d-glucose provides a rapid, efficient, and reliable semi-quantitative method. It is was effectively screened out approximately 3000 mutants for sophorolipids and the utilization of metabolizing glycerol and glucose.

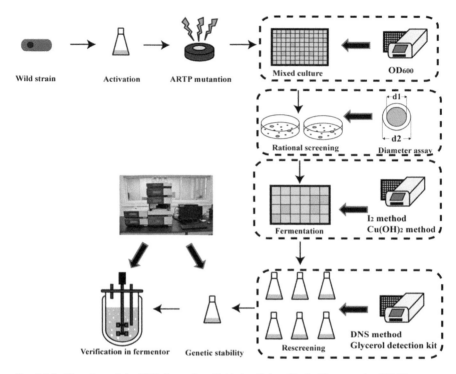

Fig. 2.19 Flowchart of the HTS for sophorolipids by *C. bombicola* (Courtesy by: [104])

Limitations Co-estimation of the carbon source and sophorolipids has a chance of human error and the complete quantification.

2.8.6 Target-Site Directed Rational HTS for High Sophorolipids

The present approach of the screening also relies on ARTP mutagenesis and an HTS system developed earlier [106]. Besides, ARTP and sodium nitrite mutagenesis have been used together with a relaxation culture (Fig. 2.20). The malonic and iodoacetic acid was used as a selective pressure to get high-yielding isolates. In the conventional model, the process of mutagenesis and preliminary screening is followed by the selection using a pressure plate [106]. The cell viability is effectively reduced after mutagenesis, and the mutants generated are exposed to high selection pressure in the agar plates [104]. In the relaxation method, in disparity, mutants are cultivated in stress-free conditions after mutagenesis, possibly a growing screening success chance.

Advantages High selective pressure increases the chance to get high-yielding strains.

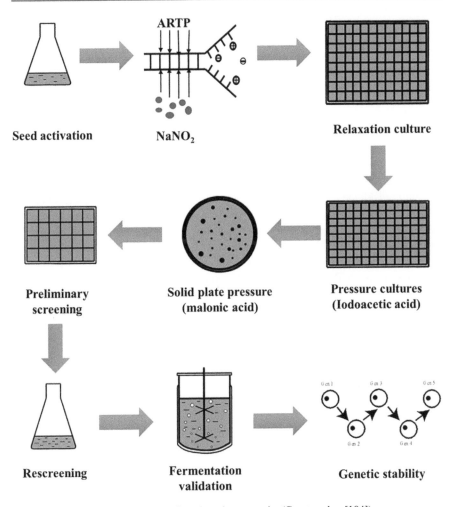

Fig. 2.20 HTS screening procedure based on the target site (Courtesy by: [104])

Limitations Therefore high-yielding mutants may be undetected due to slow growth or lethal stresses.

2.8.7 Metagenomic Approach

Microbial surfactant producers have been screened and obtained from bacteria, archaea, and eukaryotes, which the majority of the time isolated from environmental samples. Retrieving the producers from unsampled sites that may contain an unexplored population of new BSs producers is an approach to avoid the repetitive isolation of previously known producers [107]. It is well known that approximately

99% of microorganisms in any environmental site are not acquiescent to culture [108].

Unvaryingly, restrictive discovery of new BSs to conventional culturing methods may be upshot in the continual re-isolation of previously discovered BSs producers, irrespective of the environment sample tried and sampled. Avoiding the need to make ideal culture methods for difficult to isolate or culture microbes, the total environmental genomic DNA in the sample, considered as metagenome, which is recovered and analyzed for BSs potential [109, 110].

Metagenomics has been not explored in its full swing for the discovery of novel BS synthetic pathways. Given the number and population of BSs of producing strains isolated by conventional methods, it is anticipated that discovering the non-cultured element of the environmental samples by exploring metagenomics can pay significantly to novel BS discoveries. Two different strategies for exploring the new BSs producers are sequence-based screening and functional-based screening.

Direct determination of the homology to known genes or BS biosynthetic pathways using PCR based and next-generation sequencing is the major step in sequence-based metagenomics. Various online genomic databases for the detection of biosynthetic pipelines exist such as antiSMASH, BiosurfDB, MBSP1 [111–113]. In functional screening, randomly digested environmental genomic DNA is cloned into a plasmid that is further cultivated in a microbial host, i.e. *Escherichia coli*. Various well demonstrated surfactant screening approaches are acquiescent to HTS [32] and can be established to execute functional screening on a huge population of clone libraries.

2.8.8   Functional Metagenomics

Functional metagenomics recognizes genes and metabolic maze in a microbial clone library through their potential functions and gives different major benefits than that of sequence-based screening, typically as no prior acquaintance of the genes towards target molecules [114]:

- BSs possess various diverse classes structurally, screening of biochemical characteristics can be anticipated to simply produce novel structures.
- Additionally, it can be predictable that screening based on function may raise the possibility of recognizing completely novel classes of BSs.
- Identification of activity characterizes the definite accomplishment of expression in the selected heterologous host, which will enable downstream processing and evaluation. Various well-recognized screening approaches that are adapted to HTS BSs activity detection are accessible to recognize a BS producing clone.

Advantages The functional-based approach is existing for novel molecules with BS activity to be revealed using metagenomics. Equal screening in various bacterial hosts permits a different metabolic climate for the heterologous construction of a secondary metabolite of a metagenomic clone library.

Limitations The bacterial artificial chromosome (BACs), cosmids, and fosmids permit large fragments of environmental genomic DNA to be cloned, which may carry whole biosynthetic pathways. If pathways are very large, such as clusters of non-ribosomal peptides (NRPS), it is not possible to model the screening.

A major procedural constrain in the functional screening of metagenomics is that majority of genes in the genomic DNA cannot be retrieved by the library host's transcriptional system.

2.8.9 Metagenome-Derived Ornithine Lipids Screening

To identify the uncultured bacteria, more than one host has been used for heterologous expressions such as in *E. coli, Pseudomonas putida,* and *Streptomyces lividans* shuttle vector [115]. The sequence investigation and transposon mutagenesis established that an ornithine acyl-ACP N-acyltransferase was accountable for the activity. Screening for BS activity in more than one vector increases the variety of sequences that can be recognized through metagenomic.

Advantages Indicating the advantages of screening metagenomic clone libraries in hosts other than *E. coli.*

2.9   Conclusion and Perspectives

Due to the growing interest in biosurfactants, a large number of methods have been developed for the screening of strains possessing the ability of biosurfactant production. As there are both advantages as well as disadvantages associated with each method, therefore for successful screening we should employ a combination of various methods. Certain screening methods can be automated and can be employed for HTS. Thus making use of different rapid screening methods and screening a large number of isolated strains will result in the isolation of many new production strains or novel biosurfactants may be found in near future. Consequently, when new production strains will become available, the economic hindrances associated with biosurfactants will be finally overcome.

References

1. Satpute SK, Banat IM, Dhakephalkar PK, Banpurkar AG, Chopade BA (2010) Biosurfactants, bioemulsifiers and exopolysaccharides from marine microorganisms. Biotechnol Adv 28 (4):436–450
2. Satpute SK, Banpurkar AG, Dhakephalkar PK, Banat IM, Chopade BA (2010) Methods for investigating biosurfactants and bioemulsifiers: a review. Crit Rev Biotechnol 30(2):127–144
3. Satpute SK, Bhuyan SS, Pardesi KR, Mujumdar SS, Dhakephalkar PK, Shete AM, Chopade BA (2010) Molecular genetics of biosurfactant synthesis in microorganisms. In: Biosurfactants. Springer, New York, pp 14–41
4. Souza EC, Vessoni-Penna TC, de Souza Oliveira RP (2014) Biosurfactant-enhanced hydrocarbon bioremediation: an overview. Int Biodeterior Biodegradation 89:88–94
5. Pacwa-Płociniczak M, Płaza GA, Piotrowska-Seget Z, Cameotra SS (2011) Environmental applications of biosurfactants: recent advances. Int J Mol Sci 12(1):633–654
6. Ron EZ, Rosenberg E (2002) Biosurfactants and oil bioremediation. Curr Opin Biotechnol 13 (3):249–252
7. Desai JD, Banat IM (1997) Microbial production of surfactants and their commercial potential. Microbiol Mol Biol Rev 61(1):47–64
8. Varjani SJ, Rana DP, Bateja S, Sharma MC, Upasani VN (2014) Screening and identification of biosurfactant (bioemulsifier) producing bacteria from crude oil contaminated sites of Gujarat, India. Int J Inno Res Sci Eng Technol 3(2)
9. Kumar P, Sharma PK, Sharma PK, Sharma D (2015) Micro-algal lipids: a potential source of biodiesel. JIPBS 2(2):135–143
10. Jarvis FG, Johnson MJ (1949) A glyco-lipide produced by *Pseudomonas aeruginosa*. J Am Chem Soc 71(12):4124–4126
11. Hošková M, Schreiberová O, Ježdík R, Chudoba J, Masák J, Sigler K, Řezanka T (2013) Characterization of rhamnolipids produced by non-pathogenic *Acinetobacter* and *Enterobacter* bacteria. Bioresour Technol 130:510–516
12. Banat IM, Franzetti A, Gandolfi I, Bestetti G, Martinotti MG, Fracchia L et al (2010) Microbial biosurfactants production, applications and future potential. Appl Microbiol Biotechnol 87 (2):427–444
13. Perfumo A, Smyth TJP, Marchant R, Banat IM (2009) Production and roles of biosurfactants and bioemulsifiers in accessing hydrophobic substrates. In: Timmis KN (ed) Microbiology of hydrocarbons, oils, lipids, and derived compounds. Springer, London. (in press)
14. Zhao F, Zhou JD, Ma F, Shi RJ, Han SQ, Zhang J, Zhang Y (2016) Simultaneous inhibition of sulfate-reducing bacteria, removal of H2S and production of rhamnolipid by recombinant *Pseudomonas stutzeri Rhl*: applications for microbial enhanced oil recovery. Bioresour Technol 207:24–30
15. Varjani SJ, Upasani VN (2017) Critical review on biosurfactant analysis, purification and characterization using rhamnolipid as a model biosurfactant. Bioresour Technol 232:389–397
16. Ron EZ, Rosenberg E (2001) Natural roles of biosurfactants: Minireview. Environ Microbiol 3 (4):229–236
17. Batista SB, Mounteer AH, Amorim FR, Totola MR (2006) Isolation and characterization of biosurfactant/bioemulsifier-producing bacteria from petroleum contaminated sites. Bioresour Technol 97(6):868–875
18. Sharma D, Dhanjal DS, Mittal B (2017) Development of edible biofilm containing cinnamon to control food-borne pathogen. Journal of Applied Pharmaceutical Science 7(01):160–164
19. Chittepu OR (2019) Isolation and characterization of biosurfactant producing bacteria from groundnut oil cake dumping site for the control of foodborne pathogens. Grain & Oil Science and Technology 2(1):15–20
20. Sharma, D., & Saharan, B. S. (Eds.). (2018). Microbial cell factories. CRC Press: Boca Raton

21. Jamal A, Qureshi MZ, Ali N, Ali MI, Hameed A (2014) Enhanced production of rhamnolipids by *Pseudomonas aeruginosa JQ927360* using response surface methodology. Asian J Chem 26(4):1044

22. Floris R, Scanu G, Fois N, Rizzo C, Malavenda R, Spanò N, Lo Giudice A (2018) Intestinal bacterial flora of Mediterranean gilthead sea bream (Sparus aurata Linnaeus) as a novel source of natural surface active compounds. Aquac Res 49(3):1262–1273

23. Joshi S, Bharucha C, Desai AJ (2008) Production of biosurfactant and antifungal compound by fermented food isolate *Bacillus subtilis 20B*. Bioresour Technol 99(11):4603–4608

24. Mohd Isa MH, Shamsudin NH, Al-Shorgani NKN, Alsharjabi FA, Kalil MS (2020) Evaluation of antibacterial potential of biosurfactant produced by surfactin-producing *Bacillus* isolated from selected Malaysian fermented foods. Food Biotechnol 34(1):1–24

25. Sharma D, Singh Saharan Bs (2014) Simultaneous production of biosurfactants and bacteriocins by probiotic *Lactobacillus casei* MRTL3. Int J Microbiol 2014

26. Sarafin Y, Donio MBS, Velmurugan S, Michaelbabu M, Citarasu T (2014) *Kocuria marina BS-15* a biosurfactant producing halophilic bacteria isolated from solar salt works in India. Saudi J Biol Sci 21(6):511–519

27. Hassanshahian M (2014) Isolation and characterization of biosurfactant producing bacteria from Persian gulf (Bushehr provenance). Mar Pollut Bull 86(1–2):361–366

28. Bodour AA, Miller-Maier RM (1998) Application of a modified drop-collapse technique for surfactant quantitation and screening of biosurfactant-producing microorganisms. J Microbiol Methods 32(3):273–280

29. Lee DW, Lee H, Kwon BO, Khim JS, Yim UH, Kim BS, Kim JJ (2018) Biosurfactant-assisted bioremediation of crude oil by indigenous bacteria isolated from Taean beach sediment. Environ Pollut 241:254–264

30. Huang XF, Liu J, Lu LJ, Wen Y, Xu JC, Yang DH, Zhou Q (2009) Evaluation of screening methods for demulsifying bacteria and characterization of lipopeptide bio-demulsifier produced by *Alcaligenes sp.* Bioresour Technol 100(3):1358–1365

31. Perfumo A, Smyth T, Marchant R, Banat I (2010) Production and roles of biosurfactants and bioemulsifiers in accessing hydrophobic substrates. In: Handbook of hydrocarbon and lipid microbiology. Springer, London, pp 1501–1512

32. Walter V, Syldatk C, Hausmann R (2010) Screening concepts for the isolation of biosurfactant producing microorganisms. In: Biosurfactants. Springer, New York, NY, pp 1–13

33. Sharma V, Garg M, Devismita T, Thakur P, Henkel M, Kumar G (2018) Preservation of microbial spoilage of food by biosurfactantbased coating. *Asian J. Pharm. Clin. Res 11*(2):98

34. Singh J, Sharma D, Kumar G, Sharma NR (eds) (2018) Microbial bioprospecting for sustainable development. Springer, Berlin

35. Abderrahmani A, Tapi A, Nateche F, Chollet M, Leclère V, Wathelet B et al (2011) Bioinformatics and molecular approaches to detect NRPS genes involved in the biosynthesis of kurstakin from Bacillus thuringiensis. Appl Microbiol Biotechnol 92(3):571–581

36. Carrillo PG, Mardaraz C, Pitta-Alvarez SI, Giulietti AM (1996) Isolation and selection of biosurfactant-producing bacteria. World J Microbiol Biotechnol 12(1):82–84

37. Siegmund I, Wagner F (1991) New method for detecting rhamnolipids excreted by *Pseudomonas* species during growth on mineral agar. Biotechnol Tech 5(4):265–268

38. Jain DK, Collins-Thompson DL, Lee H, Trevors JT (1991) A drop-collapsing test for screening surfactant-producing microorganisms. J Microbiol Methods 13(4):271–279

39. Morikawa M, Hirata Y, Imanaka T (2000) A study on the structure–function relationship of lipopeptide biosurfactants. Biochimica et Biophysica Acta (BBA)-Molecular and Cell Biology of Lipids 1488(3):211–218

40. Cooper DG, Goldenberg BG (1987) Surface-active agents from two *Bacillus* species. Appl Environ Microbiol 53(2):224–229

41. Vaux D, Cottingham M 2001 Methods and apparatus for measuring surface configuration. Patent number WO 2007/0329729A1

42. Lindahl M, Faris A, Wadström T, Hjerten S (1981) A new test based on 'salting out' to measure relative hydrophobicity of bacterial cells. Biochimica et Biophysica Acta (BBA)-General Subjects 677(3–4):471–476
43. Mozes N, Rouxhet PG (1987) Methods for measuring hydrophobicity of microorganisms. J Microbiol Methods 6(2):99–112
44. Mnif I, Ghribi D (2015) Microbial derived surface active compounds: properties and screening concept. World J Microbiol Biotechnol 31(7):1001–1020
45. Burch AY, Shimada BK, Browne PJ, Lindow SE (2010) Novel high-throughput detection method to assess bacterial surfactant production. Appl Environ Microbiol 76(16):5363–5372
46. Rotenberg Y, Boruvka L, Neumann A (1983) Determination of surface tension and contact angle from the shapes of axisymmetric fluid interfaces. J Colloid Interface Sci 93(1):169–183
47. Van der Vegt W, Van der Mei HC, Noordmans J, Busscher HJ (1991) Assessment of bacterial biosurfactant production through axisymmetric drop shape analysis by profile. Appl Microbiol Biotechnol 35(6):766–770
48. Bakshi MS (1993) Micelle formation by anionic and cationic surfactants in binary aqueous solvents. J Chem Soc Faraday Trans 89:4323–4326
49. Banat I, Thavasi R, Jayalakshmi S (2011) Biosurfactants from marine bacterial isolates. In: Current research, technology and education topics in applied microbiology and microbial biotechnology. Formatex Research Center, Badajoz, pp 1367–1373
50. Dwivedi A, Kumar A, Bhat JL (2019) Production and characterization of biosurfactant from Corynebacterium species and its effect on the growth of petroleum degrading bacteria. Microbiology 88(1):87–93
51. Tuleva B, Christova N, Cohen R, Stoev G, Stoineva I (2008) Production and structural elucidation of trehalose tetraesters (biosurfactants) from a novel alkanothrophic Rhodococcus wratislaviensis strain. J Appl Microbiol 104(6):1703–1710
52. Singh MJ, Sedhuraman P (2015) Biosurfactant, polythene, plastic, and diesel biodegradation activity of endophytic Nocardiopsis sp. mrinalini9 isolated from Hibiscus rosasinensis leaves. Bioresour Bioproc 2(1):2
53. Kim CH, Lee DW, Heo YM, Lee H, Yoo Y, Kim GH, Kim JJ (2019) Desorption and solubilization of anthracene by a rhamnolipid biosurfactant from Rhodococcus fascians. Water Environ Res 91(8):739–747
54. Dillon JK, Fuerst JA, Hayward AC, Davis GH (1986) A comparison of five methods for assaying bacterial hydrophobicity. J Microbiol Methods 6(1):13–19
55. Sharma D, Ansari MJ, Al-Ghamdi A, Adgaba N, Khan KA, Pruthi V, Al-Waili N (2015) Biosurfactant production by Pseudomonas aeruginosa DSVP20 isolated from petroleum hydrocarbon-contaminated soil and its physicochemical characterization. Environ Sci Pollut Res 22(22):17636–17643
56. Shoeb E, Ahmed N, Akhter J, Badar U, Siddiqui K, ANSARI F et al (2015) Screening and characterization of biosurfactant-producing bacteria isolated from the Arabian Sea coast of Karachi. Turk J Biol 39(2):210–216
57. Thavasi R, Sharma S, Jayalakshmi S (2011) Evaluation of screening methods for the isolation of biosurfactant producing marine bacteria. J Pet Environ Biotechnol S 1(2)
58. Ong SA, Wu JC (2018) A simple method for rapid screening of biosurfactant-producing strains using bromothymol blue alone. Biocatal Agric Biotechnol 16:121–125
59. Morikawa M, Ito M, Imanaka T (1992) Isolation of a new surfactin producer Bacillus pumilus A-1, and cloning and nucleotide sequence of the regulator gene, psf-1. J Ferment Bioeng 74 (5):255–261
60. Nayarisseri A, Singh P, Singh SK (2018) Screening, isolation and characterization of biosurfactant producing Bacillus subtilis strain ANSKLAB03. Bioinformation 14(6):304
61. Budsabun T (2015) Isolation of biosurfactant producing bacteria from petroleum contaminated terrestrial samples that collected in Bangkok, Thailand. Procedia Soc Behav Sci 197:1363–1366

62. Bernheimer AW, Avigad LS (1970) Nature and properties of a cytolytic agent produced by *Bacillus subtilis*. Microbiology 61(3):361–369

63. Johnson MK, Boese-Marrazzo DEBORAH (1980) Production and properties of heat-stable extracellular hemolysin from *Pseudomonas aeruginosa*. Infect Immun 29(3):1028–1033

64. Morán AC, Martinez MA, Siñeriz F (2002) Quantification of surfactin in culture supernatants by hemolytic activity. Biotechnol Lett 24(3):177–180

65. Schulz D, Passeri A, Schmidt M, Lang S, Wagner F, Wray V, Gunkel W (1991) Crude-oil degrading marine microorganisms from the North-Sea. J Biosci 46:197–203

66. Persson A, Molin G (1987) Capacity for biosurfactant production of environmental *Pseudomonas* and *Vibrionaceae* growing on carbohydrates. Appl Microbiol Biotechnol 26 (5):439–442

67. Bodour AA, Drees KP, Maier RM (2003) Distribution of biosurfactant-producing bacteria in undisturbed and contaminated arid southwestern soils. Appl Environ Microbiol 69 (6):3280–3287

68. Das K, Mukherjee AK (2007) Comparison of lipopeptide biosurfactants production by *Bacillus subtilis* strains in submerged and solid state fermentation systems using a cheap carbon source: some industrial applications of biosurfactants. Process Biochem 42 (8):1191–1199

69. Ellaiah P, Prabhakar T, Sreekanth M, Taleb AT, Raju PB, Saisha V (2002) Production of glycolipids containing biosurfactant by Pseudomonas species. Ind J Exp Biol 40 (9):1083–1086

70. Haba E, Espuny MJ, Busquets M, Manresa A (2000) Screening and production of rhamnolipids by *Pseudomonas aeruginosa 47T2* NCIB 40044 from waste frying oils. J Appl Microbiol 88(3):379–387

71. Willumsen PA, Karlson U (1996) Screening of bacteria, isolated from PAH-contaminated soils, for production of biosurfactants and bioemulsifiers. Biodegradation 7(5):415–423

72. Rosenberg E, Zuckerberg A, Rubinovitz C, Gutnick D (1979) Emulsifier of *Arthrobacter RAG-1*: isolation and emulsifying properties. Appl Environ Microbiol 37(3):402–408

73. Neu TR, Härtner T, Poralla K (1990) Surface active properties of viscosin: a peptidolipid antibiotic. Appl Microbiol Biotechnol 32(5):518–520

74. Mukherjee S, Das P, Sen R (2009) Rapid quantification of a microbial surfactant by a simple turbidometric method. J Microbiol Methods 76(1):38–42

75. Chen CY, Baker SC, Darton RC (2007) The application of a high throughput analysis method for the screening of potential biosurfactants from natural sources. J Microbiol Methods 70 (3):503–510

76. Kaczorek E, Pijanowska A, Olszanowski A (2008) Yeast and bacteria cell hydrophobicity and hydrocarbon biodegradation in the presence of natural surfactants: rhamnolipides and saponins. Bioresour Technol 99(10):4285–4291

77. Pruthi V, Cameotra SS (1997) Rapid identification of biosurfactant-producing bacterial strains using a cell surface hydrophobicity technique. Biotechnol Tech 11(9):671–674

78. Smyth CJ, Jonsson P, Olsson E, Soderlind O, Rosengren J, Hjertén S, Wadström T (1978) Differences in hydrophobic surface characteristics of porcine enteropathogenic *Escherichia coli* with or without K88 antigen as revealed by hydrophobic interaction chromatography. Infect Immun 22(2):462–472

79. Svanberg M, Westergren G, Olsson J (1984) Oral implantation in humans of Streptococcus mutans strains with different degrees of hydrophobicity. Infect Immun 43(3):817–821

80. Rosenberg M, Gutnick DL, Rosenberg E (1980) Bacterial adherence to hydrocarbons. In: Microbial adhesion to surfaces. Ellis Horwood, Chichester, pp 541–542

81. Hoorfar M, Neumann AW (2010) 3 axisymmetric drop. Appl Surf Thermodyn 151:107

82. Porob S, Nayak S, Fernandes A, Padmanabhan P, Patil BA, Meena RM, Ramaiah N (2013) PCR screening for the surfactin (sfp) gene in marine *Bacillus* strains and its molecular characterization from *Bacillus tequilensis NIOS11*. Turk J Biol 37(2):212–221

83. Kavitha PG, Jonathan EL, Nakkeeran S (2012) Effects of crude antibiotic of Bacillus subtilis on hatching of eggs and mortality of juveniles of Meloidogyne incognita. Nematol Mediterr

84. Almoneafy AA, Kakar KU, Nawaz Z, Li B, Chun-lan Y, Xie GL (2014) Tomato plant growth promotion and antibacterial related-mechanisms of four rhizobacterial *Bacillus* strains against *Ralstonia solanacearum*. Symbiosis 63(2):59–70

85. Mora I, Cabrefiga J, Montesinos E (2011) Antimicrobial peptide genes in Bacillus strains from plant environments. Int Microbiol 14(4):213–223

86. Gond SK, Bergen MS, Torres MS, White JF Jr (2015) Endophytic Bacillus spp. produce antifungal lipopeptides and induce host defence gene expression in maize. Microbiol Res 172:79–87

87. Abdallah RAB, Stedel C, Garagounis C, Nefzi A, Jabnoun-Khiareddine H, Papadopoulou KK, Daami-Remadi M (2017) Involvement of lipopeptide antibiotics and chitinase genes and induction of host defense in suppression of Fusarium wilt by endophytic *Bacillus spp.* in tomato. Crop Prot 99:45–58

88. Frikha-Gargouri O, Ben Abdallah D, Bhar I, Tounsi S (2017) Antibiosis and bmyB gene presence as prevalent traits for the selection of efficient *Bacillus* biocontrol agents against crown gall disease. Front Plant Sci 8:1363

89. Khedher SB, Boukedi H, Kilani-Feki O, Chaib I, Laarif A, Abdelkefi-Mesrati L, Tounsi S (2015) *Bacillus amyloliquefaciens AG1* biosurfactant: putative receptor diversity and histopathological effects on Tuta absoluta midgut. J Invertebr Pathol 132:42–47

90. Schmidberger A, Henkel M, Hausmann R, Schwartz T (2013) Expression of genes involved in rhamnolipid synthesis in *Pseudomonas aeruginosa PAO1* in a bioreactor cultivation. Appl Microbiol Biotechnol 97(13):5779–5791

91. Morita T, Konishi M, Fukuoka T, Imura T, Kitamoto D (2006) Discovery of Pseudozyma rugulosa NBRC 10877 as a novel producer of the glycolipid biosurfactants, mannosylerythritol lipids, based on rDNA sequence. Appl Microbiol Biotechnol 73(2):305–313

92. Hsieh FC, Li MC, Lin TC, Kao SS (2004) Rapid detection and characterization of surfactin-producing *Bacillus subtilis* and closely related species based on PCR. Curr Microbiol 49 (3):186–191

93. Ottenheim C, Nawrath M, Wu JC (2018) Microbial mutagenesis by atmospheric and room-temperature plasma (ARTP): the latest development. Bioresour Bioproc 5(1):12

94. Zhang X, Zhang X, Xu G, Zhang X, Shi J, Xu Z (2018) Integration of ARTP mutagenesis with biosensor-mediated high-throughput screening to improve L-serine yield in *Corynebacterium glutamicum*. Appl Microbiol Biotechnol 102(14):5939–5951

95. Tan J, Chu J, Hao Y, Guo Y, Zhuang Y, Zhang S (2013) High-throughput system for screening of cephalosporin C high-yield strain by 48-deep-well microtiter plates. Appl Biochem Biotechnol 169(5):1683–1695

96. Tian X, Zhou G, Wang W, Zhang M, Hang H, Mohsin A et al (2018) Application of 8-parallel micro-bioreactor system with non-invasive optical pH and DO biosensor in high-throughput screening of l-lactic acid producing strain. Bioresour Bioproc 5(1):20

97. Lv X, Song J, Yu B, Liu H, Li C, Zhuang Y, Wang Y (2016) High-throughput system for screening of high l-lactic acid-productivity strains in deep-well microtiter plates. Bioprocess Biosyst Eng 39(11):1737–1747

98. Tan J, Chu J, Wang Y, Zhuang Y, Zhang S (2014) High-throughput system for screening of *Monascus purpureus* high-yield strain in pigment production. Bioresour Bioproc 1(1):16

99. Yang H, Yu H, Shen Z (2015) A novel high-throughput and quantitative method based on visible color shifts for screening *Bacillus subtilis THY*-15 for surfactin production. J Ind Microbiol Biotechnol 42(8):1139–1147

100. Vioque E (1984) Spray reagents for thin-layer chromatography (TLC) and paper chromatography (PC). Handbook of chromatography Lipids 2:309–317

101. Satake K, Rasmussen PS, Luck JM (1960) Arginine peptides obtained from thymus histone fractions after partial hydrolysis with *Streptomyces griseus* proteinase. J Biol Chem 235 (10):2801–2809

102. Potisatityuenyong A, Rojanathanes R, Tumcharern G, Sukwattanasinitt M (2008) Electronic absorption spectroscopy probed side-chain movement in chromic transitions of polydiacetylene vesicles. Langmuir 24(9):4461–4463

103. Zhu L, Xu Q, Jiang L, Huang H, Li S (2014) Polydiacetylene-based high-throughput screen for surfactin producing strains of *Bacillus subtilis*. PLoS One 9(2):e88207

104. Lin Y, Chen Y, Li Q, Tian X, Chu J (2019) Rational high-throughput screening system for high sophorolipids production in *Candida bombicola* by co-utilizing glycerol and glucose capacity. Bioresources and Bioprocessing 6(1):17

105. Norkus E, Vaškelis A, Vaitkus R, Reklaitis J (1995) On cu (II) complex formation with saccharose and glycerol in alkaline solutions. J Inorg Biochem 60(4):299–302

106. Zhou G, Tian X, Lin Y, Zhang S, Chu J (2019) Rational high-throughput system for screening of high sophorolipids-producing strains of *Candida bombicola*. Bioprocess Biosyst Eng 42 (4):575–582

107. Rizzo C, Michaud L, Hörmann B, Gerçe B, Syldatk C, Hausmann R et al (2013) Bacteria associated with sabellids (Polychaeta: Annelida) as a novel source of surface active compounds. Mar Pollut Bull 70(1–2):125–133

108. Rappé MS, Giovannoni SJ (2003) The uncultured microbial majority. Annual Reviews in Microbiology 57(1):369–394

109. Handelsman J, Liles M, Mann D, Riesenfeld C, Goodman RM (2002) Cloning the metagenome: culture-independent access to thediversity and functions of the uncultivated microbial world. Methods Microbiol 33:241–255

110. Jackson SA, Borchert E, O'Gara F, Dobson AD (2015) Metagenomics for the discovery of novel biosurfactants of environmental interest from marine ecosystems. Curr Opin Biotechnol 33:176–182

111. Weber T, Blin K, Duddela S, Krug D, Kim HU, Bruccoleri R et al (2015) antiSMASH 3.0—a comprehensive resource for the genome mining of biosynthetic gene clusters. Nucleic Acids Res 43(W1):W237–W243

112. de Oliveira MR, Magri A, Baldo C, Camilios-Neto D, Minucelli T, Celligoi MAPC (2015) Sophorolipids a promising biosurfactant and it's applications. Int J Adv Biotechnol Res 6 (2):161–174

113. da Silva Araújo SC, Silva-Portela RC, de Lima DC, da Fonsêca MMB, Araújo WJ, da Silva UB et al (2020) MBSP1: a biosurfactant protein derived from a metagenomic library with activity in oil degradation. Sci Rep 10(1):1–13

114. Tuffin M, Andaerson D, Heath C, Cowan DA (2009) Metagenomic gene discovery: how far have we moved into novel sequence space? Biotechnol J Healthcare Nutr Technol 4 (12):1671–1683

115. Williams W, Kunorozva L, Klaiber I, Henkel M, Pfannstiel J, Van Zyl LJ et al (2019) Novel metagenome-derived ornithine lipids identified by functional screening for biosurfactants. Appl Microbiol Biotechnol 103(11):4429–4441

Commercial Production, Optimization, and Purification

3

Abstract

Microbial surfactants have achieved a significant space in diverse applications in environmental clean-ups, sequestering contaminants such as hydrocarbons, and soil contaminated with heavy metals, as stable food additives, and pharmaceutical formulations because of their biodegradability, stability to the extreme environmental conditions. Although, industrial production of BSs has not been attained stipulated demand due to their low yields and production cost incurred during the bioprocess. The present book chapter took the opportunity to elaborate on the existing experimental strategies that have been implemented to make the BS production cost-effective which includes utilization of the inexpensive agro-industrial residues optimized and effective bioprocesses and hyper-producer strains developed using metabolic engineering and recombinant tools. The necessities for next-generation novel hyper-producer BS producing strains should be under the following lines, extension of substrate spectrum, increased metabolic flux specific to the substrates, low-level of by-products generation, and better bioprocess control. Recombinant/mutant strain with enhanced yields may be a tool to decrease substrate price and improve productivities sufficiently so that in due course a point is attained, where commercial production of BS develops economically viable. Soon, BS may be efficiently on a large scale as greener amphiphiles.

Keywords

Low-cost substrates · Optimization of the process · Commercialization of surfactants · Downstream processing

© Springer Nature Singapore Pte Ltd. 2021
D. Sharma, *Biosurfactants: Greener Surface Active Agents for Sustainable Future*,
https://doi.org/10.1007/978-981-16-2705-7_3

3.1   Introduction to the Commercial Production of BSs

BSs have attained a significant place in diverse applications in microbial enhanced oil recovery, removal of contaminants such as hydrocarbons, and heavy metals from affected environmental sites, as food additives, and pharmaceutical preparations [1–4] due to their effective biodegradability, stability to the harsh environmental factors such as temperature variations, pH, and ionic concentration. The commercial production of BSs has not been achieved due to their low titer and cost incurred in upstream and downstream processing. The present book chapter took the opportunity to elaborate on the existing experimental strategies that have been implemented to make the BS production cost-effective which include: utilization of the inexpensive agro-industrial optimized and effective bioprocesses and hyper-producer strains developed using metabolic engineering and recombinant tools. Sometimes control of cell signaling molecules, i.e. quorum sensing also leads to the higher production of BSs such as in the case of rhamnolipids production using *Pseudomonas aeruginosa* [5]. *P. aeruginosa* carries two interconnected QS systems, viz. the *las* and *rhl* that regulate various steps together with rhamnolipid expression, required enzyme production, and pigment production [6].

BSs can also be synthesized by recombinant engineered strains for diverse applications assuming that there is a sufficient understanding of the cell genetics makeup of the cell. It has been established and demonstrated in the case of surfactin obtained from *Bacillus subtilis*, with recent reports relating to the development of novel peptides from the genetically recombinant of different peptide synthetases. The recent use of dynamic metabolic engineering approaches for controlled gene expression could reduce the cost of bioprocess by increasing the yield of BSs. The demonstration and reports of such approaches explicitly the optimization of the cultural conditions, new hyper-producing strains, and optimization of bioprocess in a bioreactor, might lead on the way to the efficient commercial production of microbial amphiphiles in coming years.

During the past decade, the BS synthesis by different strains has been reported extensively, and enough information related to their structure, composition, and characteristics. Regardless of their commercial competitiveness and interested characters as compared to their chemical counterparts, the production, and utilization of BSs at an industrial scale has not been established due to their low concentrations and high operational cost including cost incurred in upstream and downstream processing. The BS production economy is majorly governed by three elementary factors [7]:

1. Cost of the raw substrate, i.e. low-cost industrial waste or agricultural residue.
2. Sustainable production and minimal resource-effective recovery procedures.
3. The BSs concentration is achieved by the producer strain. Thus, considering the economic limitations associated with BS production, different approaches were implemented globally to make bioprocess inexpensive and competitive:
   • The utilization of low-cost residues to decrease the initial cost involved in media preparation in the process;

- Optimization of the cultural conditions and inexpensive upstream and downstream process for maximal product recovery; and
- Development of hyper-producing mutant or modified strains for overproduction of BSs yields.

The present book chapter focuses on diverse reports, with special importance on the development and utilization of low-cost substrates, development of the overproducing strains, different types of bioprocess developed, and separation techniques and also attempts to specify the direction to take on the way to their commercial production.

3.2 Need and Availability of Inexpensive Substrates

The selection of low cost substrate for the production of BSs is the most vital factor. It is well evident that most of the biotechnological bioprocesses require high economic inputs and obtaining a maximal yield of product at minimal expenses through the utilization of the low-cost substrates [8]. BSs production economy is the key bottleneck aspect in the majority of the bioprocess. A very low yield of BSs is typically obtained from the producer and the recovery costs lies somewhere ~60–80% of the total operational cost. Therefore, most of the commercial products based on BS and BE are quite exclusive.

Every so often, the quantity and type of a substrate can contribute substantially to the production expenses; it is roughly predicted that the substrate cost may account for 10 to 30% of the overall production expenses in BSs bioprocess. Therefore, it is vital to decrease the production expenses of BSs using cost-effective and renewable raw materials [9]. One option utilized broadly is the utilization of inexpensive and agricultural residues as raw materials for BS production.

The utilization of alternative raw materials like agricultural residues is one of the striking approaches for economical BS production. Kosaric [10] proposed the utilization of agro-industrial or domestic wastewaters, rich in carbon and nitrogen nutrients, to obtain multiple advantages of diminishing the pollutants and production of BSs simultaneously. The major problem associated with low-cost substrates is the selection of appropriate waste with the precise balance of carbon and nitrogen that allows significant growth and product formation. Another approach for decreasing the reducing operational costs is creating processes that use inexpensive raw substrates or high-pollutant wastes. An extensive range of alternative renewable sources are presently available as substrates for industrial fermentations, viz. agro-industrial, and industrial waste or by-products [11, 12]. Millions of tons of industrial waste under the tag of hazardous and non-hazardous substrates are accumulated every year across the globe. It is a huge economic burden for the agro-industrial sector for the disposal of waste. Thus, efficient and appropriate utilization and management is required to manage agro-industrial wastes via the impression: reduce, reuse, and recycle. It has been recommended that efficient strategies for better economical production technologies of BS will be a combined strategy involving bioprocess development and supplies of renewable substrates [13].

A whole range of carbon and nitrogen sources have been explored for BSs production which may subsequently differ in structure or accumulation of BS either intracellular, cell-bound, or extracellular depending upon the substrate constituents predominantly the carbon source [14]. Different solid-state fermentation can be used for the economic production of BSs using cost-effective low-cost substrates at the industrial scale [15].

Ideally, the selection and utilization of the low-cost substrates mainly centered on the availability, complexity of the tropical agro-industrial crops, and residues [7, 16]. Low-cost agro-industrial substrates comprise crops like cassava [17], soybean waste and oil [18], sugar beet [19], potato waste [20], and orange fruit peel [21], agricultural crop residues such as wheat straw [22]; soy hull [23]; sugarcane bagasse [24]; fruit waste residues [25].

Furthermore, additional substrates have been advised for BS production, particularly water-miscible substrates such as molasses [26], whey milk [27], or distillery wastes [28].

3.3   Possible Substrates for BSs Production: Status Quo

3.3.1   Advantages and Drawbacks of Low-Cost Residues for BS Production

In the last decade, BSs achieved more and more interest, since large-scale production bioprocesses turn out to be more considerable, which display the potential to substitute chemical surface-active agents. Although, the application of BSs is limited to a level and segments, as the production bioprocess is not viable as compared to the chemically synthesized surfactants obtained from petrochemicals. It is due to the high price of nutrient-rich substrates, comparatively low product concentrations, and difficulty in product recovery.

While there is a whole range of nutrient substrates which have been explored to generate rhamnolipids, which are ranging from agro-industrial waste to petrochemical associated substrates, such as agricultural residues, oil processing by-products and, sugars rich substrates and glycerol. It has been reported that various waste residues may also be utilized to produce rhamnolipids, like fatty acids [29], waste frying and olive oil [18, 30], cheese whey discard [31]. In near future, waste material may turn out to be more significant, as they are typically low-cost, and they are not directly in the competition of food and substrates, and it has capitalized on the usage efficiency concerning the overall production bioprocess. Furthermore, utilizing waste nutrient substrates for microbial fermentation may also demonstrate to be advantageous for environmental solutions.

It was Mulligan and Gibbs [32] who has mentioned that a cost estimation of up to 50% of the total bioprocess cost can be considered for carbon and nitrogen sources. The typical production of rhamnolipid biosurfactants in batch fermentations has been documented to yield 0.1–0.62 g rhamnolipids/g medium (YP/S). Consequently, inexpensive substrates such as agricultural residues are one approach, secondly enhancing the BS concentrations attained vividly affects the production expenses of BSs.

Which agricultural residues nutrient mixtures to end finds their utilization for mass production of BSs will be decided within a bioprocess considering ecological, economic, and political characteristics. The local availability of such substrates and their composition along with the cost is an important aspect. Sometimes the cost of transportation and logistics costs higher in the case of substrates available in high dilutions. In such cases, high gravity substrates such as molasses, obtained from sugarcane and soy molasses have clear advantages and it is accessible around the globe almost anywhere.

3.3.1.1 Industrial Waste Valorization

Modern society generates huge quantities of waste matter through industrial activity, agro-industrial practices, and municipalities' waste [33, 34]. Legal framework and environmental guidelines to decrease the waste build-up and to promote waste valorization to boost encourage, reuse, recycling, and generation of microbial metabolites from waste biomass. Various agro-industrial activities are the source of various sources of lignocellulosic and food waste materials [16, 35].

The valorization of the industrial and agricultural residues plays an important role in commercial production as well as elimination of the waste also. The present book chapter is a compilation of reports which explored the production of BSs using various substrates obtained chiefly from renewable agro-industrial products. In place of disposing and discarding the industrial and agricultural wastes, the current inclination of research is focused on exploiting it as an economically reasonable energy source [36]. The richness of sugars, growth factors, vitamins, and micronutrients, marks them as an efficient substrate for BSs production [37–39]. A range of inexpensive agro-industrial by-products and nutrient-rich industrial wastes such as plant-derived sugar and residual oil, pulp or hydrolysates, refinery waste, dairy industry waste, potato starchy material, cheese whey, and distillery spent wash and effluents, etc., have been documented to support BS production (Table 3.1).

3.3.1.2 Industrial Waste Rich in Sugar Composition

The production of BSs, e.g. using carbohydrates or sugar-rich residues has been explored in various reports as a substrate [1, 8, 60]. Various sugar-rich waste has been generated and accumulated every year from different sources. The food processing sector generates a range of sugar-containing waste spent streams, most prominently produced by sugar processing industries such as molasses, fruit and kitchen waste, potato, starchy, cassava waste, and dairy industry cheese whey [61]. Sugarcane or beet molasses are industrial by-products which is a rich source of sucrose, glucose, and fructose. Molasses are semi-liquid residues obtained from the sugar-producing unit, which are obtained during the crystallization of the sugar syrup of sugarcane and sugarbeet. Liquid sugarcane molasses typically contain approximately 50% of sugars accessible for bioprocess (sucrose). The typical sugars

Table 3.1 Different agricultural residues/feedstocks to produce biosurfactants, microorganisms involved and obtained yield

Feedstock's	Characteristics	Organisms & Yield	References
Industrial by-products/waste			
Olives mill wastewater (OMW)	OMW is a black color liquor containing sugar, nitrogenous compounds, and residual olive oil.	*Bacillus subtilis* DSM 3256 (248.5 surfactin mg/L)	Maass et al. [40]
Sunflower soap stocks	Soap stock is gummy amber-colored rich in fatty acids	*Pseudomonas aeruginosa* (15.9 rhamnolipids g/L)	Benincasa et al. [30]
Waste frying oil	Waste frying oil comprises of various fatty acids such as palmitic acid, stearic acid, and oleic acid	*Pseudomonas aeruginosa* mutant EBN-8 (rhamnolipid 2.7 g/g)	Raza et al. [41]
Cassava starch-rich waste	Cassava waste contains about 40–45% residual starch	*Bacillus subtilis* ATCC 21332 (surfactant 2.2 g/L)	Nitschke and Pastore [17]
Cheese whey waste	Cheese whey is a rich dairy by-product contains up to 5% lactose rich in nitrogenous content	*Pseudomonas aeruginosa* strain, SR17 (rhamnolipids 4.8 g/L)	Patowary et al. [42]
Distillery waste	Distillery waste is a source of valuable substances such as polyphenols, polysaccharides, and volatile fatty acids	*Pseudomonas aeruginosa* strain BS2	Dubey and Juwarkar [43]
Palm fatty acid distillate (PFAD)	The refining process of crude palm oil produces a by-product called palm fatty acid distillate (PFAD) that is low value, renewable, and abundant waste substrate.	*Pseudomonas aeruginosa* PAO1 (rhamnolipids 3.4 g/L)	Radzuan et al. [44]
Oil refinery waste	Residual fatty acids obtained from post-refinery	*Candida antarctica* (13.24 g/L)	Bednarski et al. [45]
Waste office paper hydrolysate	Lignocellulosic pulp	*Bacillus velezensis* ASN1 (0.818 g/L)	Nair et al. [46]
Bakery waste	Bakery waste includes dough, flour dust, burnt and broken biscuits, burnt or rejected bread, and market-returned expired bakery products.	*Pseudomonas aeruginosa* strain PG1 (rhamnolipids 11.56 g/L)	Patowary et al. [47]
Agriculture crop residues			
Natural cashew apple juice	Natural cashew apple juice is an example of an inexpensive substrate, a by-product of the cashew nut industry which is rich in starch, amino acids, and micronutrients such as iron and calcium	*Acinetobacter calcoaceticus*	Rocha et al. [48]

(continued)

Table 3.1 (continued)

Feedstock's	Characteristics	Organisms & Yield	References
Soy molasses	Soy molasses is produced from the alcoholic extraction of the sugars (cellulose & other sugars) and nitrogenous compounds present in the defatted bran.	*Pseudomonas aeruginosa* ATCC 10145 (11.7 rhamnolipids g/L)	Rodrigues et al. [49]
Sugarcane molasses	Molasses contain high sugar content (30–50%) along with other nitrogen, phosphorous and minor elements.	*Bacillus subtilis* RSL-2 (12.34 surfactin g/L)	Verma et al. [50]
Ground-nut oil refinery residue	Ground-nut oil refinery residue is rich in fatty acids (60%), carbohydrates (35), and various micronutrients such as sodium, magnesium, potassium, and zinc.	*Candida lipolytica* UCP 0988 (4.5 g/L)	Rufino et al. [51]
Sunflower oil	Rich source of fatty acids	*Pseudomonas aeruginosa* PAO1 (rhamnolipids 39 g/L)	Mueller et al. [52]
Safflower oil	Rich source of fatty acids	*Pseudomonas aeruginosa* GS9–119 (rhamnolipids 2.98 g/L)	Rahman et al. [53]
Canola oil	Co-utilization of canola oil and glucose	*Candida lipolytica* (8 g/L)	Sarubbo et al. [54]
Sugarcane bagasse hemicellulosic hydrolysate	Lignocellulosic residues are rich in carbohydrates contents such as cellulose and hemicellulose, and their hydrolyzed fractions are rich in glucose, arabinose, and xylose	*Cutaneotrichosporon mucoides* UFMG-CM-Y6148 ($0.167 \text{ g/L}^{-1}.\text{h}^{-1}$)	Marcelino et al. [24]
Rapeseed oil	–	*Pseudomonas* species DSM 2874 (rhamnolipids 45 g/L)	Trummler et al. [55]
Turkish corn oil	–	*Candida bombicola* ATCC 22214 (Sophorolipids 400 g/L)	Pekin et al. [56]
Barly pulp	Barley pulp is a rich source of glucose, fructose, and various other monosaccharides	*Pseudomonas aeruginosa* ATCC 9027 (rhamnolipids 9.3 g/L)	Kaskatepe et al. [57]
Orange peel		*Pseudomonas aeruginosa* MTCC 2297 (rhamnolipids 9.18 g/L)	George and Jayachandran [21]
Lignocellulosic residues	Rich in xylose and other simpler sugars	*Achromobacter* sp. (PS1) (4.13 g/L)	Joy et al. [58]
Sunflower acid oil	Fatty acids	*Pseudomonas aeruginosa* (rhamnolipids 4.9 g/L)	Jadhav et al. [59]

present in molasses are sucrose (~29% of total carbohydrates), glucose (~12%), and fructose (~13%). Molasses contain approximately 22% water and about 75% carbohydrates, and no added protein or fat. The typical choice of molasses as a substrate for BSs production is the widespread availability and its low price as compared to other sugary substrates, molasses is also rich in various vitamins and growth factors [62]. It was demonstrated by Joshi et al. [26] that molasses and cheese whey can be used for the production of BS production by *Bacillus subtilis*. Same way, soy molasses is a by-product obtained during soybean processing plant in a huge quantity with low commercial value [49]. Soy molasses is an agricultural residue generates in high volume with low commercial value. Soy molasses has huge potential in fermentative modeling due to the richness of carbohydrates, lipids, and fractions of proteins. Soy grains also have a significant amount of carbohydrates, predominantly present as cellulose and sugar, possessing a significant fraction of nitrogenous substances [63]. Soy molasses is rich in sugar and obtained from the alcoholic extraction of the carbohydrates present in the defatted soy bran [64]. Rodrigues et al. [49] explored the soy molasses as a low-cost substrate for the production of the rhamnolipids using *Pseudomonas aeruginosa* ATCC 10145. It was observed that after 48 h of fermentation a concentration of 11.70 g/L was attained. It was established from the above-mentioned results that it is possible to obtain rhamnolipids from *Pseudomonas aeruginosa* 10,145 using soybean molasses and can be a potential carbon-rich substrate.

Similarly, the cheese whey obtained from the dairy sector has also been documented as an inexpensive and potentially viable substrate for BS production. Cheese whey is a lactose-rich by-product obtained during cheese production, and it is also containing approximately 12–14% proteins [9]. Cheese whey has a high biological oxygen demand (BOD) value and its throwing away can be challenging particularly for countries depending on the dairy economy. It has been observed that approximately 50% of the cheese whey generated annually is valorized into useful products like food additives and active ingredients, the rest is considered as a wastewater stream.

The cheese whey supports the growth of various BS producing strain growth and is utilized as a cost-effective substrate for BS production. The conversion of dairy lactose for fermentation developments has been a point of consideration for various researchers in the last decades. Though only a few bacteria can breakdown galactose from the lactose moiety, the accumulation of residual galactose is one serious concern, the complete metabolization of the galactose may not be attained without modified/engineered strains.

The production of glycolipid BS (rhamnolipids) by *Pseudomonas aeruginosa* using cheese whey has been demonstrated by Dubey and Juwarkar [31] and it was found to be a 1 g/L yield. The production of the Sophorolipids by yeast using cheese whey concentrate and rapeseed oil has been demonstrated by Daniel et al. [65]. It was established that a yield of 422 g/L attained in a sequential fermentor. First, deproteinized cheese whey concentrate was utilized for the growth of *Cryptococcus curvatus* ATCC 20509. In the sequential stage, yeast biomass generated from the first stage was homogenized for sophorolipid production. The sophorolipids production by *Candida bombicola* in a medium containing cheese whey with glucose, oleic acid, and yeast extract. The concentration of the 34 g/L was obtained and achieved under optimized concentrations [66].

A large amount of sugar-rich wastewater from industrial origin can be potentially utilized as a raw material for the fermentation medium [8]. Distillery spent wash is obtained in huge volume with high BOD value, highly problematic for environmental discharge. The spent wash typically comprises varying ratios of organic and inorganic acidic matter, with high-intensity color with an unpleasant odor. The *Pseudomonas aeruginosa* was cultivated to produce rhamnolipids using distillery spent wash as a nutrient source [67]. Food processing and agro-industrial residues could all help as possible substrates for the production of BSs.

Different waste residues containing starchy biomass are also conventional substitute substrates for the production of BSs. Potato waste or effluent of potato processing units is a potential residue for BSs production is available from the potato processing industry. Potato waste or potato effluents generate billions of pounds, and only approximately 59% of the potato material is processed to the finished products [68]. Potato processing industry effluent is discharging as a starch-rich stream that can be harmful to environmental bodies. The majority of the starchy waste is an economic burden to the manufacturers and environmental bodies; consequently, manufacturers have sought innovative approaches to get rid of their waste. Generally, the potato industry waste has been utilized as animal food and feed or as a raw material for ethanol production [69].

Typically, potatoes handling unit waste contains approximately 80% water, 17% sugar-rich material, 2% protein fractions, 0.1% fat moieties, and 0.9% vitamins, and micronutrients, and trace elements [70]. Potato starch waste can be utilized by the bacteria equipped with the amylase enzymes and therefore have the huge potential to be used as a carbon source for BSs production.

Potato processing units (effluents obtained from potato processing industries) were utilized to obtain BS by *Bacillus subtilis* [20, 71–73]. Other starchy crops such as cassava, wastewater an important agricultural residue, have been broadly used as raw substrates for different agri-industrial production to obtain starch, ethanol, and other value-added chemicals. Cassava complex cellular composition with high carbon content gives them a huge potential for microbial conversion into biobased products through biorefinery which is ultimately gaining economic and environmental stability to cassava industries. Various reports confirmed that the cassava wastewater can be used as a potential substrate for the BS (surfactin) production by *Bacillus subtilis* [17, 74–76].

Nitschke and Pastore [17] found that natural cassava wastewater is a potential substrate in which the surface tension was reduced of fermentative medium to 26 mN/m by Bacillus sp. 23. Subsequently, *Bacillus subtilis* LB5a was also cultivated using the cassava wastewater and BS exhibited high surface tension decrease, low CMC value, and thermal stability, portentous potential industrial applications. The production of the BS substances on a pilot scale was also established to be an economically viable bioprocess [77].

Various other starchy waste residues, such as rice processing industries, a leftover of domestic cooking and community kitchens [78], corn steep liquor [79], and wastewater from the processing of food [80], have great potential to cultivate microbial growth and BS production.

Separately from the directly above other residues can also be utilized as a carbon source for the production of BS such as fruit peels, bakery waste, wood hydrolysate, grape marc, shrimp shell waste, etc. Patowary et al. [47] have documented bakery leftover enriched with minerals salts for the rhamnolipids productions in a concentration of 11.56 g/L using *Pseudomonas aeruginosa* PG1. Shrimp waste is also documented for the production of the 4–6 g/L rhamnolipids using *Pseudomonas stutzeri* L1 [81]. A significant yield of rhamnolipids by *Pseudomonas aeruginosa* SS14 i.e. 14.87 g/L was attained using rice grains solubles as a carbon source [82].

In a study, *Lactobacillus pentosus* was cultivated on the hydrolyzed grape marc for BS production [83]. The *Lactobacillus pentosus* was reported to produce 0.6 mg/g intracellular BS of sugar consumed. In one more study, Portilla-Rivera et al. [84] also reported the BS production using hydrolyzed grape marc with significant surface and emulsification properties. Widespread research is desirable to create the appropriateness of these carbohydrate-rich residues in the commercial level BS production.

3.3.1.3 Industrial Waste Rich in Oil, Fatty Acids, and Fats

BSs can be obtained by utilizing industrial wastes such as waste frying oils, oil industry waste, and petroleum oil residues as low-cost substrates [85]. It was demonstrated in various reports with plant-associated oils have display that they can be used as an appropriate and low-cost substrate for BS production. Various vegetable oils have been utilized for rhamnolipids production in past, e.g. soybean, sunflower, canola oil, rapeseed, palm, fish oil, olive oil, and coconut oil. Separately a diverse range of vegetable oils, oil by-products from vegetable oil processing were also documented as efficient substrates for BSs production. Additionally, oil lards, soap stocks, palm oil mill waste, olive mill waste, and different other fatty acids can be used for microbial growth and BS production. The disposal of oil processing industry waste is an environmental hazard and leads to economic loss, as oil industry wastes show low degradability because of lipid moieties present in wastewater [18].

It has been observed that rhamnolipids production by *Pseudomonas aeruginosa* utilizing soap stocks of sunflower oil soapstock as a nutrient source and attained a yield of 15.9 g/L [30]. Soap stocks originated from the various oil processing sites such as cottonseed, corn, and soybean were also reported for the rhamnolipids productions and a yield of 11.7 g/L of rhamnolipids [86]. A huge volume of the vegetable processing of waste frying oils is generated by food processing sectors. Most of the time waste cooking oil can only be used for biodiesel production or can be used as animal feed. But it has been established in various reports that waste cooking oil can be used for the efficient production of the BS such as rhamnolipids production by *Pseudomonas aeruginosa* utilizing waste frying oil [87, 88].

It is well established in the *Pseudomonas aeruginosa* that plant oil can be used as an appropriate substrate to produce rhamnolipids and the final yield of the BSs is higher than the other sugar-based substrates. However, such food processing oils and by-products are easily affordable and accessible in huge amounts across the globe. But the oils explored to produce BS production are typically edible oils and are not cost-effective, and there is a food vs substrate contradiction. On the other hand, various plant-associated oils such as jatropha, castor, babassu, and ramtil oil, etc. are not appropriate for food consumption because of their unacceptability in terms of color, smell, and compositions therefore, accessible at a lower cost. Additionally, fatty acids, which can be attained for approximately 1/10th the cost in lower purities, are also available and can be a choice of substrate for BSs production. The utilization of such low-cost oils and fatty acid wastes in the commercial fermentation media might significantly decrease the overall price of BS production, resulting in them achievable targets for forthcoming R&D expansions.

3.3.2 Advantages and Limitations of Using Agricultural Residues and Waste Effluents

The BSs production using oils and sugars residues is a well documented bioprocess. Agricultural residues/feedstocks may be utilized for animal feed and food supplementation and consequently, compete with food issues. The utilization of agricultural waste and industrial stream to produce rhamnolipids also observe various drawbacks and constraints (Table 3.2). All the agro-industrial residues are not rich in all the nutrients and inorganic salts to support the growth of BSs producing strain, complexity of the substrates requires different sets of metabolic pathways and cell pipelines to metabolize it. The availability of low-cost substrates is typically local and limited and lacks large volumes or adequate quantities to endure a commercial production of BSs. Also, the key substrate concentration within the waste material is relatively low (e.g. approximately 5% lactose in cheese whey leftover), which may further requirement of preparing concentrate. Concentrating the waste also leads to additional costs of operation and production. The diluted substrate is further raising the cost of logistics and transport.

Moreover, agro-industrial residues are composed of highly inconstant components as it depends upon the various biotic and abiotic factors, which leads to the major variability in every lot of such potential feedstocks. Although, close control and streamline bioprocess regulation can be possible to overcome such problems.

Furthermore, the political influence on agricultural practices and policy also leads to the actual problems such as palm tree plantation policy in southeast Asia is a having a space of doubts for the biobased production of BS using palm oil. So, the biobased substrates utilization in the BS production process hinders the sustainability of the approach.

Table 3.2 Advantages and constraints of agro-industrial substrates in BS production (Modified and adopted from: [8])

Advantages	Drawbacks
Industrial production cost can be minimized	Substrates composed of unwanted compounds
Low-cost agricultural residues	Compete with food and feed issues
Much cheaper/renewable	Processing or pre-treatment of the residue is a pre-requisite to obtaining fermentable carbon, nitrogen source
Substrates are available in large volumes	Final substrates or waste stream itself get impurities, color, the smell from the processing of substrates (e.g. molasses)
Agro residues are a rich source of various carbon and nitrogen moieties	All the producer's strain is not equipped with the required set of enzymatic systems to hydrolyze the substrate.
Enhanced the concentrations of BS/bio-emulsifiers	Downstream processing is difficult to get the maximal and pure product, this raises the production cost and purity afterward
Availability of the substrates, seasonal accessibility	Sometimes substrates are only available regionally, transport cost and logistics is the huge problem. Seasonal availability is also a huge problem, as the availability and cost vary in the offseason
Basic functional characteristics of the product would not change	The continuous stock of substrates with uniform constituents may not be possible
Food industries generate a large volume of waste substrates	Key substrate concentration is low despite huge volume such as cheese whey, i.e. only contains 5% (v/v) lactose concentrations
Does not contains harmful toxic contents to microorganisms	Toxicity concept may vary from substrate to substrate such as furfural accumulation during lignocellulosic waste pre-treatment affects the growth of microorganisms
Substrates are rich in various carbon, nitrogen, and vitamins	Accounts for main inconsistency in various batches of such potential feedstocks

3.4   Bioprocess for the Commercial Production of BSs

BSs are produced extracellularly and cell-associated by a diverse range of microorganisms in the stationary growth phase under nutrient limiting conditions [2]. More than 200 patents were listed before 2011 linked with the production of BSs, with 85% of such listed patents are on glycolipids including rhamnolipids and sophorolipids, 50%, and 35%, respectively [89]. The production of BSs can either be induced or natural due to the presence of various substances, such as rate of aeration, pH, inoculum size, temperature, nutrient uptake capacity, and agitation speed. The yield of BSs affected due to the composition and number of constituents such as carbohydrates, nitrogenous material, and other elements [61, 90]. A leading strategy adopted for attaining increased yields in the BSs production is by medium and bioprocess optimization.

An industrially competent and economically viable optimized bioprocess is the underpinning platform for any BS producing industry; therefore, optimization of a bioprocess is the key developmental step in the direction of commercialization. Some works related to initial kinetics and bioprocess development of BSs have been established in a few research demonstrations which fix the benchmark for commercial production. Although still there are various bottlenecks in the scale-up of the BSs production process, which should be addressed and established with appropriate statistical and computational tools.

Any effort to upsurge the concentration of a BS needs optimal inclusion of media constituents and choice of the optimal culture circumstances that will encourage the maximal or the maximal productivity. Currently, few BSs have been utilized on an industrial scale due to the lack of cost-effective production bioprocesses. Various applications that need a high volume of low-cost BS were hampered by unsuitability in the usage of such compounds. Different approaches have been anticipated and adopted to make the BS bioprocess further profitable comprises: (1) Development of efficient bioprocesses, includes optimization of medium, (2) Development of overproducing mutants and strains, (3) Rapid, low-cost downstream processing [7, 8, 41, 91] (Fig. 3.1).

In the same way, efficient downstream processing methodologies and tools are needed for maximal product recovery. Downstream processing or recovery accounts for the vital cost of biotechnology and product formation [7].

3.5   Improvement/Optimization of Bioprocess

The production of microbial surface-active agents has relied on rate-limiting substrates, such as carbon and nitrogen compositions. The importance of the nutritional supplies for BS has been discussed in the present section. Commercial production of BSs appeared to be promising, effective but so far approaches to overcome the affordability with their chemical surfactants. Commercial BSs production costs can be minimized using the optimization of process kinetics and substrate utilization. The production of BSs categorized as, growth-associated, mixed growth, non-growth associated production [17] based on the relationship between growth, substrate use, and BSs production. Various elements, nutrient constituents, media formulation, and precursors are documented to disturb the process of BSs production and the structure, and yields. The selection of carbohydrates lays a significant role in the concentration and structure of BSs. Choice of the carbon source often alters the structure of the BSs and responsible for their inhabited properties either. The solubility of the carbon source, i.e. water-soluble sugars and nitrogen source or hydrophobic oil waste determines the accumulation of BSs either its extracellular or intracellular/cell-associated [7].

Limitation or controlled availability of the various elements like iron, nitrogen, and manganese are documented to distress the yield of BSs. [92]. It has been broadly

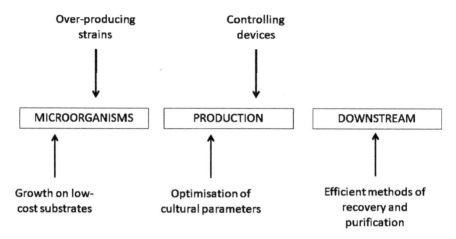

Fig. 3.1 Different cost-reduction approaches for the production of BSs (Courtesy by: [1])

reported in *Pseudomonas aeruginosa* with a strategy to overproduce the BSs in
nitrogen and iron limiting conditions. It has been documented that the overproduction of rhamnolipids by *Pseudomonas aeruginosa* under the phosphorus and sulfur
limiting conditions [93]. Nitrogenous material supply is also regarded as critical
nutrient compositions, the suitable nitrogen source for BSs production using *Bacillus
subtilis* S499 was found to be glutamic acid, valine, lysine, and alanine [94]. Likewise, the supplementation of iron and manganese to the fermentation medium was
documented to increase the BS yield accumulated by *Bacillus subtilis* [95].

The selection of low-cost substrates is available at present possible which can be
utilized as a carbon source, particularly agro-industrial residues such as sugarcane or
soy molasses, corn steep liquor, and waste originated from the food processing
industry. By utilizing such cost-effective raw materials, the bioprocess costs can
be managed for the market requirements. The effect of different carbon residues to
produce BSs using *Pseudomonas aeruginosa* sources such as n-hexadecane, olive
oil, and glucose. It was observed that the production kinetics and yield are much
better in the medium contains n-hexadecane and olive oil as a major carbon source
[96]. The effect of carbon source is also observed in the *Candida antarctica* KCTC
7804, yielded a maximal BS yield of 41.0 g L^{-1} attained by supplementing glycerol
and olive oil in initial rounds of the log phase of the fermentation kinetics [97]. The
supplementation of precursor to the fermentative media along with the carbon source
will also encourage the BS production [98]. A yield of the 31.0 g/L of the
sophorolipids was achieved by the *Candida antarctica* KCTC 7804 when
supplemented with the vegetable oil as a precursor during the process [97]. The
supplementation of fermentative media with vegetable oil resulted in an increased
value of 70.0 g/L of BSs [99].

A suitable strategy approach used for attaining higher yields in the fermentation is
by medium and bioprocess optimization. Maximizing productivity or reducing

production expenses demands the utilization of bioprocess-optimization approaches which involve multiple reasons. The conventional strategy of substrate optimization comprises shifting one variable at a given time while putting the other variables at fixed levels; but this approach is painstaking, time-consuming, and does not promise the fortitude of the maximal BS production [100].

The enhanced BS production yield has been additionally explored using a better innovative statistical tool that takes into consideration the interaction between different factors in control in the fermentation process.

To handle such issues and make the optimization process uniform, a set of statistical modeling, artificial intelligence, and computer-assisted control of BS production such as response surface methodology (RSM) has been adopted by different investigators. Various other methods such as Plackett-Burman Design [101], Taguchi Experimental Design [102], Fractional Factorial Design [103] are commonly adopted statistical tools for bioprocess development.

The RSM includes a group of statistical approaches for experimental designs, model assessments, simultaneously determining the major effects on the variables and demonstrating optimum conditions [104]. RSM explored different regression studies by using a quantitative data set got from the significantly composed determination of multivalent factors conditions simultaneously [105]. The RSM approach was efficaciously used to determine the optimum substrate concentration, inoculum, and environmental factors for the improved and higher production of BS by *Bacillus subtilis* [100, 106].

RSM approach has been significantly used employed to increase BSs yield by decreasing the production cost through identifying a balanced ratio of culture nutrients compositions and process parameters [61, 104]. Such optimization strategies would help the BS industry to model the best suitable media containing cost-effective substrates and to explore the most encouraging process parameter for enhanced production.

3.5.1 Development of Overproducing Mutants and Strains

There are four basic strategies in any biotechnological process to reduce the bioprocess cost [107].

- The strain (selection, adaptation, or engineered as hyper-producers),
- The bioprocess (optimized for low-cost substrates and operating expenses),
- The microbial growth nutrients or bioprocess feedstock,
- The process value-added products (minimum or valorization of waste to saleable products).

Even enhanced and optimized fermentation process engineering has been explored, commercial production of BSs often relies mainly on hyper-producing producer strain to develop commercially feasible products. The readiness of hyper-producer microorganisms is vital to economize the production bioprocess and to obtain BS and other by-products with improved commercial properties.

There is various strain recognized to produce BSs, such as *Pseudomonas aeruginosa, Bacillus subtilis, Candida bombicola, and Candida antarctica*. Though, by now, some handful of microbial strains are effectively relevant for commercial production method, due to the huge variances in BS concentrations and maximum attainable yields. The production of rhamnolipids by *Pseudomonas aeruginosa* is well documented possibly due to the genome of the extensively used strain PAO1 is fully annotated and available to the public domain. Though *Pseudomonas aeruginosa* is known as an opportunistic human pathogen and therefore this strain would typically not be the foremost selection if it comes to commercial production.

The genetics of the producer strain is a significant factor affecting the yield of all bioprocess due to the potential to produce BS is conferred by the genetic makeup of the producer organism. Moreover, the commercial BS production bioprocess is reliant on the readiness of recombinant or mutant hyper-producer if significant yields are deficient from the wild strains. If wild and hyper-producer strain is available, even the recombinant/mutant hyper-producer is still in need, to economize further the production bioprocess and to get metabolites with improved commercial importance. Besides, the natural BS producer strains, more hyper-producer mutant and recombinant strains with high yield are documented reported in the literature.

The option of the generation of mutant strain with N-methyl-N-nitro-N-nitrosoguanidine to rhamnolipids production by *Pseudomonas aeruginosa* has been documented [108]. The yield of the rhamnolipids was maximal, i.e. 12.5 g/L as compared to the wild strain, i.e. 1.2 g/L on the fifth and seventh day of incubation, respectively. To daze the problematic opportunistic pathogen *Pseudomonas aeruginosa*, recombinant strains of *Pseudomonas putida* and *Pseudomonas fluorescens* were obtained to produce significant yields of the rhamnolipids [109]. Recombinant microbial strains such as these can be effectively utilized in biotechnology industries where microbial safety is an important attention.

Along with the mutant type of BS producer's generation, a genetic recombination is also an important tool for genetic modification of the strain. Ohno et al. [110] developed a surfactin producer strain, Bacillus subtilis harboring a plasmid containing genes for surfactin production. A yield of the 8–50 g/L surfactin was attained during fermentation with soybeans flour-based medium.

Same time, strain improvement by mutation or recombination tools to enhance BS production is still instigating another condition due to its necessity of sophisticated tools and resources consequently, cumulative rising cost in search of the new strain. Though, such practices are unavoidable and vital to enhance the commercial BS production bioprocess and make it more feasible to make production bioprocess cost-effective.

To date, whatever research and developments have been carried out in BS production epitomize just the upper layers, and the generation and utilization of the mutant/recombinant hyper-producer strain have enormous concealed potential to enhance yield and economically viable production of BS bioprocess. On different fronts such as low-cost substrates, recovery, and the option of using hyper-producer strain are still the real revolutions, where such strains can enhance the yield many times. In near future, the BS domain can assume the development of various non-pathogenic, GRAS, potent, and hyper-producing mutants/recombinant strains (Table 3.3).

Though the potential of BSs production using low-cost agricultural and industrial substrates is by far not completely utilized, and custom-made strains purposely to produce BSs have still in their infancy. The necessities for next-generation BSs producing hyper-producer strains is still the primary goal of the BS production as follows: extension of substrate range, enhanced metabolic spectrum, low level of the by-product's accumulations, and better-controlled instruction. Strain-engineering/mutant development may be an approach to reduce substrate price and improve productivities adequate so that in the end a point is touched, where commercial production of BSs becomes economically viable.

3.5.2 Downstream Processing of BSs

Biosurfactants are not likely to be obtained at a low cost using a high-class downstream bioprocess. Thus, the downstream process progresses are desired to emphasize on BSs that offer themselves to the simple and cost-effective product recovery methods. Product recovery methods in various biotechnological bioprocesses are accounted for up to 60% of total bioprocess expenses. Because of economic attention, most BS would have to include either whole-cell spent media broths or other crude formulations. BS recovery relied majorly on its ionic charge, water mixing, and cellular location (intracellular, extracellular, or cell-associated).

Process development for large-scale production of biosurfactants is essential to lessen the cost of raw material, processing, and recovery process. Economic approaches, which underline the utilization of waste effluents as no-cost residues and low-cost in situ product recovery approaches for recovering BSs are vital for creating a large-scale BS production technology.

More development and research are needed to maximize and optimized such existing product recovery bioprocess to make them much competitive and commercially attractive.

3.6 Production and Downstream Process

The BSs recovery/downstream processing and purification of the amphiphiles is an expensive practice as it accounts for 60–80% of the total production cost [7]. The simple, rapid, and integrated approaches for the purification and separation techniques make BS bioprocess economically competitive [61, 119].

It has to turn out to be beneficial to explore how to replace conventional approaches with competent and cost-effective tools for downstream processing and purification. In conventional approaches, huge volumes of chemical solvents are typically utilized which ultimately increasing the production cost and certainly

Table 3.3 Mutant and recombinant BS producers strains with improved BS yields and with improved product properties

Recombinant strain	Modification	Improvement	Reference
Pseudomonas aeruginosa 59C7	Transposon induced mutant	Doubles the production of BS	Koch et al. [111]
Pseudomonas aeruginosa PTCC 1637	Random mutations	Production increases 10 times	Tahzibi et al. [108]
Bacillus licheniformis KGL11	Random mutations	12 times more production	Lin et al. [112]
B. subtilis ATCC 55033	Random mutations	2–4 g /l crude surfactin	US patent no. 5227294
Pseudomonas aeruginosa EBN-8	Gamma-ray induced mutant	Increase production twice	Iqbal et al. [113]
Bacillus subtilis Suf-1	Ultraviolet mutant	Increase production twice	Mulligan et al. [114]
Acinetobacter calcoaceticus RAG-1 mutants	Selection based on cationic detergent	Increase production twice	Shabtai and Gutnick [115]
Recombinant *Bacillus subtilis* MI 113	Insertion of a plasmid	Surfactin concentration increased	Ohno et al. [110]
B. subtilis SD901	Random mutagenesis	8–50 g/l surfactin yields	US patent no 7011969
Recombinant *Bacillus subtilis* strain ATCC 21332	Inserted with a peptide synthetase using recombination	Production of lipohexapeptide	Symmank et al. [116]
Recombinant *Bacillus subtilis*	Produced by whole enzyme module swapping	Production of lichenysin	Yakimov et al. [117]
Recombinant *Pseudomonas aeruginosa* strains	Incorporation of *E. coli* lacZY genes into	To utilize lactose- and whey-based low-cost substrates	Koch et al. [118]
Recombinant *Pseudomonas putida* KT2442 and *P. fluorescens*	Heterologous expression of rhlAB genes	Production of rhamnolipids in non-pathogenic cells	Ochsner et al. [109]

the overall environmental pollution. Furthermore, conventional production and recovery methods mainly carried out in batches which reduce BS yields because of product inhibition or low concentrations.

The BSs recovery and purification is mainly depends upon the various factors such as cellular location (accumulated intracellularly or secreted extracellularly, and cell associated), charge of BS (chromatography), solubility (water/organic solvents), molecular size (ultrafiltration), foaming capacity (foam fractionation) directly influence the recovery and purification methods [120] (Table 3.4).

Crude or partially purified BSs such as glycolipids have been separated using the acid precipitation approach, which is promptly available, non-tedious and cost-effective. Such approaches have also been adopted to recover lipopeptide and

Table 3.4 Biosurfactants recovery based on their properties, methods, and relative advantages

Process	BS properties	BS type	Advantages	Laboratory resources required
Acid precipitation	BS turn out to be insoluble at low range pH	Surfactin, rhamnolipids & other glycolipids	Low-cost, simple, rapid	No experimental pre-requisite required
Solvent extraction	BS hydrophobic moiety is soluble in organic solvent	Trehalolipids, Sophorolipids, Liposan	Efficient, fast, reusable in nature	No experimental pre-requisite required
Ammonium sulfate precipitation	Salting out of the protein-rich fractions	Lipopeptides and Glycolipopeptides	Low-cost, batch mode, more efficient than acid precipitation	No experimental pre-requisite required
Crystallization	Gravity separation	Glycolipids	Batch process, efficient	–
Centrifugation	BS separated based on their insoluble properties	Glycolipids	Reusable, fast for crude recovery	High speed centrifugation with a provision of refrigeration required
Adsorption	Adsorbed on wood charcoal and polystyrene and can be desorbed with organic solvents	Rhamnolipids, lipopeptides, and other glycolipids	Efficient, continuous, high purity, regeneration possibilities	Glass columns, polystyrene resins
Foam fractionations	Foam forming capacity	Surfactin and glycolipopeptides	Continuous mode, efficient, high purity	Special extension and modification in bioreactor head plate required
Tangential flow filtration	Molecular size	Mixed biosurfactants	Continuous process, fast, efficient, can be scaled up	Tangential flow filtration systems required
Ion exchange chromatography	Ionic charge	Glycolipids	Fast recovery, high purity, reusability	Ion exchange resins required
Membrane ultrafiltration	Micelle formation by biosurfactants	Glycolipids and surfactin	Simple, fast, requires no phase change, no chemical added	Ultrafiltration unit with desired molecular cut-offs (hollow fibers and molecular cassettes)

(continued)

Table 3.4 (continued)

Process	BS properties	BS type	Advantages	Laboratory resources required
Acetone precipitation	Precipitation	Bio-emulsifiers	Simple and efficient	No experimental pre-requisite required
Methyl tertiary butyl ether (MTBE) solvent extraction	Hydrophobic moiety of BS is soluble in MTBE	Glycolipids	Less toxic, low-cost and reusable	No experimental pre-requisite required

rhamnolipid from fermentation medium [121]. On the contrary, different organic solvents such as MTBE, chloroform, diethyl ether, methanol, ethyl acetate, and *n*-hexane are also mostly used approaches for separation of BS [119]. Filtration is managed for separation of insoluble debris from extracted constituents which is further evaporated by means of rotary evaporator [119]. BSs recovery by acid or acetone precipitation was also documented by a few reports [104].

Solvent extraction systems have used to purify diverse range of BSs viz. rhamnolipids, surfactin, sophorolipids, trehalolipids, cellobiolipids, and liposan, and trehalose lipids [7, 61]. Moreover, other low-cost methods such as ammonium sulfate precipitation, adsorption and centrifugation, membrane separation have been documented [122]. Every so often, a multifaceted downstream processing that uses a sequence of absorption and purification phases is much more efficient than a solo recovery process to get purified BS [123]. During most of the downstream processing, there should be proper attention due to contamination risk with undesired metabolites.

The above-mentioned approaches have foremost drawbacks including;

- High cost or expensive due to the utilization of huge volumes of organic solvents and time constraints,
- Production of a large number of organic solvents as toxic leftover leads to disposal problems,
- Loss of BS potential due to the utilization of organic solvents, and
- Final-product degradation as an outcome of product aggregation in batch fermentations.

3.7 Integrated Separation System of BSs

Various constraints and prospects are existing in BS production because of the surface-active properties, predominantly foam formation, and membrane separations. Certain contests give rise to the prospect to grow and apply various integrated tools with efficiency for BS recovery. In this section, an overview of the present constraints faced in BS secreted out of the cell and converses how innovative integrated product recovery strategies can handle such issues.

Every so often, a single downstream processing method is not enough for efficient product recovery and purification. The in situ downstream strategies permit continuous recovery of BSs from fermentation medium leading to an enhancement in yield and system efficiency. For BSs mass production, improvement of production efficiency of the recovery methods and the rate of BS removal is critical. Integrated approach collectively recovery methods and production strategies as rapid and effective practices have been growing.

Integrated production and recovery of BSs are nowadays also being analyzed for various bioprocesses where mass transfer challenges and restrictions and the need to separate production and product recovery are a vital thought. A range of such methods has been documented for the integrated product recovery approach, with adsorption, solvent extraction, membrane separations, gas stripping, and foam fractionation.

3.7.1 Foam Fractionation

Foam fractionation, a recently used method for previously identified has been regarded as an inexpensive and eco-friendly approach that is appropriate for the recovery of BS from a large volume of fermentation media. The air bubbles produced with an aeration structure, move up to the surface of the fermentation media leading to the development of the foam fraction. A firm amount of liquid is captured between air bubbles or foam lamella and is so lost within the foam.

Various BSs of industrial interest will collect quickly at an intentionally developed air–water interface, i.e. an air bubble interface, consequential in the development of a stable foam in an extremely agitated, bubbly fermenter. If such foaming is unrestrained foam will collect the fermenter headspace and, in the modest foam fractionation process, a BS containing foam runoffs from the fermenter via the air vent in an unrestrained way and is removed in an overflow decanter.

Foam fractionation has been explored as an integrated product recovery approach for various BSs such as surfactin, rhamnolipids, and other glycolipids [124]. Foam fractionation technology has various advantages and attained a specific interest to recover the BSs from bioreactor at industrial production (Fig. 3.2). Foam fractionation is documented for the separation of surfactin, rhamnolipids, and other protein-rich surfactants such as hydrophobin proteins, and its most effective with a dilute yield of BSs [126].

A well-designed product recovery method for in situ rhamnolipids production and purification is foam fractionation. It was well demonstrated previously that strain cultivation and integrated product recovery is an efficient method for BSs production [120, 127, 128]. But, production of rhamnolipids and integrated foam fractionation was revealed to be not possible due to high amount of cellular biomass in the foam leading to the inevitability to develop resolutions concerning cell retention such as connection of magnetic separation or cell biomass recycling [129]. Integrated

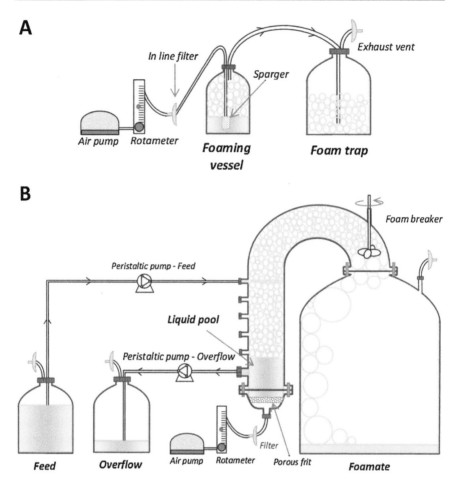

Fig. 3.2 Diagrams of the experimental at shake flask-scale batch foaming (**a**) and foam fractionation in bioreactor (**b**) (Courtesy by: [125])

production and foam fractionation during rhamnolipids production by heterologous production *Pseudomonas putida* KT2440 strain is documented with low cell biomass enrichment in the foamate providing the method to remove highly concentrated glycolipid from the fermentation broth in situ [126] (Fig. 3.3).

3.7.2 Membrane Separation

In addition to the other integrated production and separation methods for BS research, membrane separation depends upon the selective or pre-fixed penetrability of a used membrane to separate the amphiphile of curiosity in a production broth with a pressure gradient to maintain flow across the membrane. Microfiltration cut

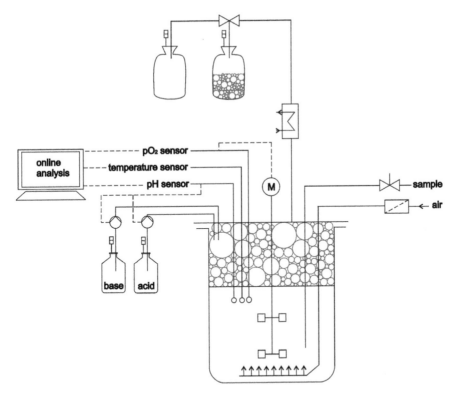

Fig. 3.3 Schematic diagrams for integrated foam fractionation in a fermenter for the separation of rhamnolipids (Courtesy by: [126])

off membrane with desired pore size can be utilized to primarily recover biomass which is flowing back to the fermenter in the retentate, while the BS rich and biomass free permeate is percolated to an ultrafiltration surface membrane. The retentate of the suspension has the maximum BS whereas the water and other molecules are permeable to the ultrafiltration membrane. The membrane-based recovery has been explored for the surfactin, yeast sophorolipids, and other glycolipids [130, 131]. Such reports have mostly been achieved at laboratory scale, there are only a handful of such attempts has been documented at pilot scale, but Ecover industries have tried to scale-up the method to pilot plant capacity scale [132]. The membrane separation method has the benefit of demonstrating the fermentation and recovery to be completely decoupled, after primary recovery with microfiltration, which can effortlessly be applied with soluble BSs [133].

An ultrafiltration membrane depended integrated production and recovery system was used for sophorolipids, which are predominantly highly soluble at >500 gL^{-1}, to selectively separate cells, with an additional membrane used to recover the product. Such modification and integrated production and separation method allow a rise in BS productivity ranges from 0.37 gL^{-1} h^{-1} to 0.63 gL^{-1} h^{-1} [134]. Coutte et al. [130] demonstrated that membrane separation can be used for the separation of surfactin, with increasing the surfactin product concentration. Integrated production, cell recycling, and product separation enables productivity to a total of approximately 110 $mgL^{-1} h^{-1}$ and improved the total surfactin production to 10 g from 3 L fermentation batch.

Several such integrated production and separation methods, foam fractionation, membrane recovery, ultrafiltration, and gravity-based separation, have been demonstrated to be efficient at downstream processing of the amphiphiles. Such integrated production and separation methods like a membrane, gravity-based recovery, and foam fractionation have been used to significantly improve bioprocess economics, attaining the attainable product yields and overall process productivity. Application of such methods in industrial production and at scale will allow us the reduced operational cost of BSs, and consequently an immensely extended BSs market.

There are certain challenges associated with the integrated production and separations also such as the harsh recovery situations that are frequently performed, along with the unsuitability of most of the separating instruments utilized with cell biomass, because of biofouling, biofilm development, and microbial contamination. Adsorption columns or separation membranes are particularly developed with biofilms are difficult in cleaning and charging. Such constraints in a cell removal system before product recovery can occur and leads to making such methods excessively difficult and expensive.

3.7.3   Metabolic Engineering of Strain for Higher Production

There are various approaches used to enhance the rhamnolipid production by suitable strains. Primarily, the producing strain may be culturally optimized to cultivate an extensive range of nutrients which cannot ideally be utilized by the strain (e.g. by incorporating genes responsible for proteins for transporters and/or metabolic enzymes having the potential to metabolize the substrate into exploitable transition metabolites) thereby introductory cheaper residues for the production of BS and improving the litheness of the bioprocess. Subsequently focus should be on the optimization of microbial metabolic pathways, by predicting optimized cell pipelines for catabolism and production of rhamnolipids. Consequently, the approach should be engrossed in decreasing by-products and precursors for such metabolic pathways, which enticement substrates from the pipelines of BS synthesis. The last element is engrossed in regulation and induction of BS synthesis, e.g. by heterologous products of genes associated with BS synthesis under inducible promoters or by uncoupling rhamnolipid production from various cellular regulation, such as quorum sensing independent rhamnolipid productions [135].

3.7.4 Commercialized Biosurfactants

In recent days, the commercial production of the BSs has attained a special market and demand. Biosurfactant production by the *Bacillus subtilis, Candida bombicola,* and certain strains of *Pseudomonas aeruginosa* have been utilized for commercial production. Strains of the Pseudomonas sp. are reported for the rhamnolipids production but associated with the human pathogenesis. As the strain safety in the bioprocessing turns out to be paramount for commercial production with the search for novel non-pathogenic BSs producers. Surface-active agents obtained from probiotic organisms will play a critical role in the future of the food surfactants market. Rhamnolipid could be a possible substitute for the chemical surfactant molecules and critical platform with the market potential of $2.8 billion in 2023. There is an instant requirement for the search of non-pathogenic BSs producers with improved production capacity and exertions to scale-up using bioprocess optimization are imperative to meet the future estimates of the BSs market. Various industries have been optimized and producing BSs at a commercial scale using different wild and heterologous strains (Table 3.5).

3.7.5 Biosurfactant Production Economics

There is growing awareness among the global population now a days with concern to sustainability, green solutions, and global warming. The rising demand for biobased surfactants is ever increasing and "green chemicals or additives" are required for every process. While various researchers are enthusiastic in replacing chemical surfactants, the inflated cost of producing BSs is of prime concern. Presently, low yield and the high price of substrates delay probable inexpensive BSs production on the commercial scale [135]. Decrease in the of BSs production making them viable on economic front which make them attractive mainly relies on the screening of the novel strains, improvement of bioprocess and operational cost, utilization of low-cost feedstocks, improved bioprocess yield, media simulations and optimization, or use of recombinant and superlative mutants, with integrated separation technologies [137]. The best option of inexpensive or free substrates is of vast importance to the total economy of the process with the share of total production cost ensuing to 30–50% [138, 139]. Henkel et al. [135] reported the cost of the substrate in relation to the production of rhamnolipids (Table 3.6). Author compared the available renewable substrate to their cost incurred to produce the rhamnolipids.

Furthermore, the expenses for the BSs recovery and purification are still comparatively high. Stage by stage expense should be minimized for BSs production to a minimal rate as much as possible, such the cost of substrates like oil-rich by-products that could be valorized into BSs in large volumes and substantial yields [140]. The selection of the cost-effective feedstocks depends upon its availability in large volumes can greatly varies from region to region, such as corn steep liquor derived from the corn will be cost-effective in countries with overproducing corn, as

Table 3.5 Various industries/organizations in biosurfactants production (Adopted and modified from Randhawa and Rahman [136])

Industry	Type of BS	Applications area
TeeGene biotech	Rhamnolipids and Lipopeptides	Pharmaceuticals and cosmetics
AGAE technologies	Rhamnolipids	Pharmaceutical, cosmetics, and bioremediation
Jeneil biosurfactant co.	Rhamnolipids	Cleaning and care products, MEOR
Paradigm biomedical Inc.	Rhamnolipids	Pharmaceuticals
Rhamnolipid companies, Inc.	Rhamnolipids	Agriculture, pharmaceuticals, bioremediation, food, and, pharmaceutical
Fraunhofer IGB	Glycolipids, Cellobiose lipids, MELs	Cleansing products and personal care
Saraya co. ltd.	Sophorolipids	Cleaning products, personal care, hygiene health products
Synthezyme	Sophorolipids	Cleaning and cosmetics
Ecover, Belgium	Sophorolipids	Cleaning, cosmetics formulations, pest and mosquito control
Groupe Soliance	Sophorolipids	Cosmetics
MG Intobio Co. Ltd.	Sophorolipids	Beauty and personal care
Synthezyme	Sophorolipids	Cleaning and cosmetics
Allied carbon solutions (ACS)	Sophorolipids	Agricultural formulations, ecological applications
Henkel	Sophorolipids, Rhamnolipids, Mannosylerythritol lipids	Glass cleaning formulations, laundry, cosmetics
Kaneka Co.	Sophorose lipids	Cosmetics and toiletry products
Groupe Soliance	Sophorolipids	Cosmetics
Rhamnolipids holdings	Rhamnolipids	Personal care
Green Pyramid Ltd.	Sophorolipids	Food and postharvest control
Vetline Ltd.	Glycolipids	Poultry emulsifiers

compared to the non-corn or sugarcane growers in Asia and other developing countries.

Various other agricultural feedstocks such as sugarcane molasses and date molasses are also available for the lipopeptides production [141]. In order to contest with the chemical surfactant, the cost of the bioprocess must be brought down to 2 USD per liter which is in a challenging job. There are various factors which need to be minimized to achieve economic production of the BSs, therefore there is no shortage of the opportunities in commercial production of such amphiphiles. Co-metabolites and other by-products are value-added materials and increase the economic viability

Table 3.6 Global production rates, present market prices and theoretical substrate conversion for low-cost feedstocks potentially used for rhamnolipids production (Adopted and modified (cost changed from EURO to USD) from: [135])

Substrate	Key constituents	Global production (million tons/year)	Substrate cost (USD per kg/rhamnolipid)
Sugar cane	Glucose and fructose	140	1.02
Molasses	Glucose and fructose	51	0.64
Lignocelluloses	Cellulose	500	0.20
Cheese whey	Lactose	4	2.39
Biodiesel crude	Glycerol	1	0.25
Fatty acids	Octadecanoic acid	–	0.070

of any business. The estimation and simultaneous production of the other by-products along with BSs such as enzymes is an additional advantage which can be encashed to surge the net profit. Esterase obtained from the BSs production from various recombinant or mutant strain will open the new initiative to reduce the production cost [137, 142].

The use of value-added, even if from a separate bioprocess could be additional smart solution such as glycerol, which is a co-metabolite of biodiesel production. Glycerol is accessible in surplus volume in the global market as a low-cost feedstock for BSs production [143]. BSs research needs to be intensive on suitable vigorous production microorganisms, inexpensive substrates, and economic viable bioreactor technology. The present market cost of rhamnolipid (R-95, 95%) is USD 227/10 mg (Sigma-aldrich) and USD 200/10 mg (AGAE technologies, USA) calling for determined research.

The average constituents of the agricultural feedstocks may lead to high price for gravity and/or purification of highly diluted chief constituents, as it can predominately be found in food processing industry waste. Such feedstocks may also responsible for higher logistic costs. To overcome such drawbacks, we should consider the lignocellulosic feedstocks and hydrolysates as a prominent feedstock for BSs production.

3.7.6 Research Requirements and Future Directions

Microbial surfactants production from waste agro-industrial waste feedstocks would be a favorable strategy for resource recovery. The research developments to be carried out for allowing low-cost & environmentally friendly production and integrated separation technologies. BSs can be produced by diverse strains using various industrial substrates; however, it involves further analysis for scale-up to produce such amphiphiles at a large scale. Various bottlenecks are supposed to be fixed to support BS production and separation from agro-industrial feedstocks as an advantageous option for value-added production also. For maximal separation of the BS, efficient integrated production and separation approaches such as foam fractionation and membrane separation should be adopted. The task would be the adoption of low-cost strategies for separation which would lead to maximal BS recovery at a nominal cost.

Further research should determine different factors that may affect BS production.

- Partially purified BS can be adopted for agricultural and environmental applications such as bioremediation. The stability of the microbial surfactants should be determined under harsh environmental conditions to improve their efficacy.
- The quality and bioprocess yield of BS can be improved by utilizing hyper-producing mutants/recombinant cells.
- Novel methods and strategies are essential to enhance yield and reduce production costs.
- BS production by utilizing wastes as a substrate needs to be demonstrated at a commercial scale as a biorefinery concept to support the waste valorization method.

Though the potential of BS production using various feedstocks is fully utilized, and producer strains specifically tailored to obtain BS have yet to be established. The necessities for next-generation novel hyper producer BS producing strains should be under the following directions; extension of substrate spectrum increased metabolic flux specific to the substrates, low- the level of by-products generation, and better bioprocess control. Recombinant/mutant strain with enhanced yields may be a tool to decrease substrate price and improve productivities sufficiently so that in due course a point is attained, where commercial production of BS develops economically viable. Soon, BS may be efficiently on a large scale as greener amphiphiles.

References

1. Banat IM, Franzetti A, Gandolfi I, Bestetti G, Martinotti MG, Fracchia L et al (2010) Microbial biosurfactants production, applications and future potential. Appl Microbiol Biotechnol 87 (2):427–444
2. Sharma D, Saharan BS, Kapil S (2016) Biosurfactants of lactic acid bacteria. Springer, Cham
3. Kumar P, Sharma PK, Sharma PK, Sharma D (2015) Micro-algal lipids: a potential source of biodiesel. J Innov Pharm Biol Sci 2(2):135–143
4. Sharma V, Sharma D (2018) Microbial biosurfactants: future active food ingredients. In: Microbial bioprospecting for sustainable development. Springer, Singapore, pp 265–276
5. Dusane DH, Zinjarde SS, Venugopalan VP, Mclean RJ, Weber MM, Rahman PK (2010) Quorum sensing: implications on rhamnolipid biosurfactant production. Biotechnol Genet Eng Rev 27(1):159–184
6. Smith RS, Iglewski BH (2003) P. aeruginosa quorum-sensing systems and virulence. Curr Opin Microbiol 6(1):56–60

7. Mukherjee S, Das P, Sen R (2006) Towards commercial production of microbial surfactants. Trends Biotechnol 24(11):509–515

8. Banat IM, Satpute SK, Cameotra SS, Patil R, Nyayanit NV (2014) Cost effective technologies and renewable substrates for biosurfactants' production. Front Microbiol 5:697

9. Makkar RS, Cameotra SS, Banat IM (2011) Advances in utilization of renewable substrates for biosurfactant production. AMB Express 1(1):5

10. Kosaric N (1992) Biosurfactants in industry. Pure Appl Chem 64(11):1731–1737

11. Da Silva AC, de Oliveira FJ, Bernardes DS, de França FP (2009) Bioremediation of marine sediments impacted by petroleum. Appl Biochem Biotechnol 153(1–3):58–66

12. Makkar R, Cameotra S (2002) An update on the use of unconventional substrates for biosurfactant production and their new applications. Appl Microbiol Biotechnol 58 (4):428–434

13. Smyth TJP, Perfumo A, McClean S, Marchant R, Banat IM (2010) Handbook of hydrocarbon and lipid microbiology, vol 1. Springer, London, pp 1–6

14. Fiechter A (1992) Integrated systems for biosurfactant synthesis. Pure Appl Chem 64 (11):1739–1743

15. Kapadia SG, Yagnik BN (2013) Current trend and potential for microbial biosurfactants. Asian J Exp Biol Sci 4(1):1–8

16. Sharma D, Saharan BS (eds) (2018) Microbial cell factories. CRC Press

17. Nitschke M, Pastore GM (2004) Biosurfactant production by Bacillus subtilis using cassava-processing effluent. Appl Biochem Biotechnol 112(3):163–172

18. Haba E, Espuny MJ, Busquets M, Manresa A (2000) Screening and production of rhamnolipids by Pseudomonas aeruginosa 47T2 NCIB 40044 from waste frying oils. J Appl Microbiol 88(3):379–387

19. Onbasli D, Aslim B (2008) Determination of antimicrobial activity and production of some metabolites by Pseudomonas aeruginosa B1 and B2 in sugar beet molasses. Afr J Biotechnol 7 (24)

20. Thompson DN, Fox SL, Bala GA (2000) Biosurfactants from potato process effluents. In: Twenty-First Symposium on Biotechnology for Fuels and Chemicals. Humana Press, Totowa, pp 917–930

21. George S, Jayachandran K (2009) Analysis of rhamnolipid biosurfactants produced through submerged fermentation using orange fruit peelings as sole carbon source. Appl Biochem Biotechnol 158(3):694–705

22. Panjiar N, Mattam AJ, Jose S, Gandham S, Velankar HR (2020) Valorization of xylose-rich hydrolysate from rice straw, an agroresidue, through biosurfactant production by the soil bacterium Serratia nematodiphila. Sci Total Environ 138933

23. Marti ME, Colonna WJ, Reznik G, Pynn M, Jarrell K, Lamsal B, Glatz CE (2015) Production of fatty-acyl-glutamate biosurfactant by Bacillus subtilis on soybean co-products. Biochem Eng J 95:48–55

24. Marcelino PRF, Peres GFD, Terán-Hilares R, Pagnocca FC, Rosa CA, Lacerda TM et al (2019) Biosurfactants production by yeasts using sugarcane bagasse hemicellulosic hydroly-sate as new sustainable alternative for lignocellulosic biorefineries. Ind Crop Prod 129:212–223

25. Archana K, Reddy KS, Parameshwar J, Bee H (2019) Isolation and characterization of sophorolipid producing yeast from fruit waste for application as antibacterial agent. Environ Sustain 2(2):107–115

26. Joshi S, Bharucha C, Jha S, Yadav S, Nerurkar A, Desai AJ (2008) Biosurfactant production using molasses and whey under thermophilic conditions. Bioresour Technol 99(1):195–199

27. Sharma D, Saharan BS, Chauhan N, Bansal A, Procha S (2014) Production and structural characterization of Lactobacillus helveticus derived biosurfactant. *The Scientific World Journal*:2014

28. Dubey KV, Juwarkar AA, Singh SK (2005) Adsorption—; desorption process using Wood-based activated carbon for recovery of biosurfactant from fermented distillery wastewater. Biotechnol Prog 21(3):860–867
29. Abalos A, Pinazo A, Infante MR, Casals M, Garcia F, Manresa A (2001) Physicochemical and antimicrobial properties of new rhamnolipids produced by pseudomonas a eruginosa AT10 from soybean oil refinery wastes. Langmuir 17(5):1367–1371
30. Benincasa M, Contiero J, Manresa MA, Moraes IO (2002) Rhamnolipid production by Pseudomonas aeruginosa LBI growing on soapstock as the sole carbon source. J Food Eng 54(4):283–288
31. Dubey K, Juwarkar A (2001) Distillery and curd whey wastes as viable alternative sources for biosurfactant production. World J Microbiol Biotechnol 17(1):61–69
32. Mulligan CN, Gibbs BF (1993) Factors influencing the economics. Biosurfact Prod Prop Applicat 159:329
33. Ashekuzzaman SM, Forrestal P, Richards K, Fenton O (2019) Dairy industry derived waste-water treatment sludge: generation, type and characterization of nutrients and metals for agricultural reuse. J Clean Prod 230:1266–1275
34. Gupta GK, Shukla P (2020) Insights into the resources generation from pulp and paper industry wastes: challenges, perspectives and innovations. Bioresour Technol 297:122496
35. Champagne P (2007) Feasibility of producing bio-ethanol from waste residues: a Canadian perspective: feasibility of producing bio-ethanol from waste residues in Canada. Resour Conserv Recycl 50(3):211–230
36. Girotto F, Alibardi L, Cossu R (2015) Food waste generation and industrial uses: a review. Waste Manag 45:32–41
37. Guerrero LA, Maas G, Hogland W (2013) Solid waste management challenges for cities in developing countries. Waste Manag 33:220–232
38. Zhang X, Dequan L (2013) Response surface analyses of rhamnolipid production by Pseudomonas aeruginosa strain with two response values. Afr J Microbiol Res 7(22):2757–2763
39. Singh R, Glick BR, Rathore D (2018) Biosurfactants as a biological tool to increase micronutrient availability in soil: a review. Pedosphere 28(2):170–189
40. Maass D, Moya Ramirez I, Garcia Roman M, Jurado Alameda E, Ulson de Souza AA, Borges Valle JA, Altmajer Vaz D (2016) Two-phase olive mill waste (alpeorujo) as carbon source for biosurfactant production. J Chem Technol Biotechnol 91(7):1990–1997
41. Raza ZA, Khalid ZM, Banat IM (2009) Characterization of rhamnolipids produced by a Pseudomonas aeruginosa mutant strain grown on waste oils. J Environ Sci Health Part A 44 (13):1367–1373
42. Patoway R, Patoway K, Kalita MC, Deka S (2016) Utilization of paneer whey waste for cost-effective production of rhamnolipid biosurfactant. Appl Biochem Biotechnol 180(3):383–399
43. Dubey K, Juwarkar A (2004) Determination of genetic basis for biosurfactant production in distillery and curd whey wastes utilizing Pseudomonas aeruginosa strain. Ind J Biotechnol 8:74–81
44. Radzuan MN, Banat IM, Winterburn J (2018) Biorefining palm oil agricultural refinery waste for added value rhamnolipid production via fermentation. Ind Crop Prod 116:64–72
45. Bednarski W, Adamczak M, Tomasik J, Płaszczyk M (2004) Application of oil refinery waste in the biosynthesis of glycolipids by yeast. Bioresour Technol 95(1):15–18
46. Nair AS, Al-Bahry S, Sivakumar N (2019) Co-production of microbial lipids and biosurfactant from waste office paper hydrolysate using a novel strain Bacillus velezensis ASN1. Biomass Convers Biorefinery 10:1–9
47. Patoway R, Patoway K, Kalita MC, Deka S (2018) Application of biosurfactant for enhancement of bioremediation process of crude oil contaminated soil. Int Biodeterior Biodegradation 129:50–60
48. Rocha MV, Oliveira AH, Souza MC, Gonçalves LR (2006) Natural cashew apple juice as fermentation medium for biosurfactant production by Acinetobacter calcoaceticus. World J Microbiol Biotechnol 22(12):1295–1299

49. Rodrigues MS, Moreira FS, Cardoso VL, de Resende MM (2017) Soy molasses as a fermentation substrate for the production of biosurfactant using Pseudomonas aeruginosa ATCC 10145. Environ Sci Pollut Res 24(22):18699–18709

50. Verma R, Sharma S, Kundu LM, Pandey LM (2020) Experimental investigation of molasses as a sole nutrient for the production of an alternative metabolite biosurfactant. J Water Proc Eng 38:101632

51. Rufino RD, Sarubbo LA, Campos-Takaki GM (2007) Enhancement of stability of biosurfactant produced by Candida lipolytica using industrial residue as substrate. World J Microbiol Biotechnol 23(5):729–734

52. Mueller A, Thijs C, Rist L, Simões-Wüst AP, Huber M, Steinhart H (2010) Trans fatty acids in human milk are an indicator of different maternal dietary sources containing trans fatty acids. Lipids 45(3):245–251

53. Rahman KSM, Rahman TJ, McClean S, Marchant R, Banat IM (2002) Rhamnolipid biosurfactant production by strains of Pseudomonas aeruginosa using low-cost raw materials. Biotechnol Prog 18(6):1277–1281

54. Sarubbo LA, Farias CB, Campos-Takaki GM (2007) Co-utilization of canola oil and glucose on the production of a surfactant by Candida lipolytica. Curr Microbiol 54(1):68–73

55. Trummler K, Effenberger F, Syldatk C (2003) An integrated microbial/enzymatic process for production of rhamnolipids and L-(+)-rhamnose from rapeseed oil with pseudomonas sp. DSM 2874. Eur J Lipid Sci Technol 105(10):563–571

56. Pekin G, Vardar-Sukan F, Kosaric N (2005) Production of sophorolipids from Candida bombicola ATCC 22214 using Turkish corn oil and honey. Eng Life Sci 5(4):357–362

57. Kaskatepe B, Yildiz S, Gumustas M, Ozkan A, S. (2017) Rhamnolipid production by pseudomonas putida IBS036 and pseudomonas pachastrellae LOS20 with using pulps. Curr Pharm Anal 13(2):138–144

58. Joy S, Rahman PK, Khare SK, Soni SR, Sharma S (2019) Statistical and sequential (fill-and-draw) approach to enhance rhamnolipid production using industrial lignocellulosic hydrolysate C6 stream from Achromobacter sp.(PS1). Bioresour Technol 288:121494

59. Jadhav JV, Anbu P, Yadav S, Pratap AP, Kale SB (2019) Sunflower acid oil-based production of Rhamnolipid using Pseudomonas aeruginosa and its application in liquid detergents. J Surfactant Deterg 22(3):463–476

60. Henkel M, Schmidberger A, Vogelbacher M, Kühnert C, Beuker J, Bernard T et al (2014) Kinetic modeling of rhamnolipid production by Pseudomonas aeruginosa PAO1 including cell density-dependent regulation. Appl Microbiol Biotechnol 98(16):7013–7025

61. Saharan BS, Sahu RK, Sharma D (2011) A review on biosurfactants: fermentation, current developments and perspectives. Genet Eng Biotechnol J 2011(1):1–14

62. Makkar RS, Cameotra SS (1997) Utilization of molasses for biosurfactant production by two Bacillus strains at thermophilic conditions. J Am Oil Chem Soc 74(7):887–889

63. Smith AK, Circle SJ (1978) Historical background. In: Soybeans: chemistry and technology, vol 1. Wsestport: The AVI publishing company, Dublin, pp 1–26

64. Johnson LA (1992) Soy proteins history, prospects in food, feed. Inform 3:429–444

65. Daniel H Je, Reuss M, Syldatk C (1998) Production of sophorolipids in high concentration from deproteinized whey and rapeseed oil in a two stage fed batch process using Candida bombicola ATCC 22214 and Cryptococcus curvatus ATCC 20509. Biotechnol Lett 20 (12):1153–1156

66. Daverey A, Pakshirajan K (2010) Kinetics of growth and enhanced sophorolipids production by Candida bombicola using a low-cost fermentative medium. Appl Biochem Biotechnol 160 (7):2090–2101

67. Babu PS, Vaidya AN, Bal AS, Kapur R, Juwarkar A, Khanna P (1996) Kinetics of biosurfactant production by Pseudomonas aeruginosa strain BS2 from industrial wastes. Biotechnol Lett 18(3):263–268

68. Idaho Potato Commission (1994) Internet site. http://www.potatoes.org/dev/product.html

69. Natu RB, Mazza G, Jadhav SJ (1991) Waste utilization. In: Potato: production, processing, and products. CRC Press, Boca Raton, pp 175–201
70. Church CF, Pennington JAT, Church HN (1985) Bowes and Church's food values of portions commonly used. Lippincott (Periodicals), Philadelphia
71. Noah KS, Bruhn DF, Bala GA (2005) Surfactin production from potato process effluent by Bacillus subtilis in a chemostat. In: Twenty-Sixth Symposium on Biotechnology for Fuels and Chemicals. Humana Press, New York, pp 465–473
72. Noah KS, Fox SL, Bruhn DF, Thompson DN, Bala GA (2002) Development of continuous surfactin production from potato process effluent by Bacillus subtilis in an airlift reactor. In: Biotechnology for fuels and chemicals. Humana Press, Totowa, pp 803–813
73. Thompson DN, Fox SL, Bala GA (2001) The effect of pretreatments on surfactin production from potato process effluent by Bacillus subtilis. In: Twenty-Second Symposium on Biotechnology for Fuels and Chemicals. Humana Press, Totowa, pp 487–501
74. Costa SG, Lépine F, Milot S, Déziel E, Nitschke M, Contiero J (2009) Cassava wastewater as a substrate for the simultaneous production of rhamnolipids and polyhydroxyalkanoates by Pseudomonas aeruginosa. J Ind Microbiol Biotechnol 36(8):1063–1072
75. Nitschke M, Pastore GM (2003) Cassava flour wastewater as a substrate for biosurfactant production. In: Biotechnology for fuels and chemicals. Humana Press, Totowa, pp 295–301
76. Nitschke M, Pastore GM (2006) Production and properties of a surfactant obtained from Bacillus subtilis grown on cassava wastewater. Bioresour Technol 97(2):336–341
77. Barros FFC, Ponezi AN, Pastore GM (2008) Production of biosurfactant by Bacillus subtilis LB5a on a pilot scale using cassava wastewater as substrate. J Ind Microbiol Biotechnol 35 (9):1071–1078
78. Oje OA, Okpashi VE, Uzor JC, Uma UO, Irogbolu AO, Onwurah IN (2016) Effect of acid and alkaline pretreatment on the production of biosurfactant from rice husk using Mucor indicus. Res J Environ Toxicol 10(1):60
79. Gudiña EJ, Fernandes EC, Rodrigues AI, Teixeira JA, Rodrigues LR (2015) Biosurfactant production by Bacillus subtilis using corn steep liquor as culture medium. Front Microbiol 6:59
80. Orts Á, Tejada M, Parrado J, Paneque P, García C, Hernández T, Gómez-Parrales I (2019) Production of biostimulants from okara through enzymatic hydrolysis and fermentation with Bacillus licheniformis: comparative effect on soil biological properties. Environ Technol 40 (16):2073–2084
81. Kadam D, Savant D (2019) Biosurfactant production from shrimp shell waste by Pseudomonas stutzeri. Ind J Geo-Marine Sci 48(9):1411–1418
82. Borah SN, Sen S, Goswami L, Bora A, Pakshirajan K, Deka S (2019) Rice based distillers dried grains with solubles as a low cost substrate for the production of a novel rhamnolipid biosurfactant having anti-biofilm activity against Candida tropicalis. Colloids Surf B: Biointerfaces 182:110358
83. Rivera OMP, Moldes AB, Torrado AM, Domínguez JM (2007) Lactic acid and biosurfactants production from hydrolyzed distilled grape marc. Process Biochem 42(6):1010–1020
84. Portilla-Rivera O, Torrado A, Domínguez JM, Moldes AB (2008) Stability and emulsifying capacity of biosurfactants obtained from lignocellulosic sources using Lactobacillus pentosus. J Agric Food Chem 56(17):8074–8080
85. Müller MM, Hausmann R (2011) Regulatory and metabolic network of rhamnolipid biosynthesis: traditional and advanced engineering towards biotechnological production. Appl Microbiol Biotechnol 91(2):251–264
86. Nitschke M, Costa SG, Contiero J (2005) Rhamnolipid surfactants: an update on the general aspects of these remarkable biomolecules. Biotechnol Prog 21(6):1593–1600
87. De Lima CJB, Ribeiro EJ, Servulo EFC, Resende MM, Cardoso VL (2009) Biosurfactant production by Pseudomonas aeruginosa grown in residual soybean oil. Appl Biochem Biotechnol 152(1):156

88. Zhu Y, Gan JJ, Zhang GL, Yao B, Zhu WJ, Meng Q (2007) Reuse of waste frying oil for production of rhamnolipids using Pseudomonas aeruginosa zju. u1M. J Zhejiang Univ Sci A 8 (9):1514–1520

89. Vecino X, Cruz JM, Moldes AB, Rodrigues LR (2017) Biosurfactants in cosmetic formulations: trends and challenges. Crit Rev Biotechnol 37(7):911–923

90. Darvishi P, Ayatollahi S, Mowla D, Niazi A (2011) Biosurfactant production under extreme environmental conditions by an efficient microbial consortium, ERCPPI-2. Colloids Surf B: Biointerfaces 84(2):292–300

91. Thavasi R, Jayalakshmi S, Balasubramanian T, Banat IM (2007) Biosurfactant production by Corynebacterium kutscheri from waste motor lubricant oil and peanut oil cake. Lett Appl Microbiol 45(6):686–691

92. Hewald S, Josephs K, Bölker M (2005) Genetic analysis of biosurfactant production in Ustilago maydis. Appl Environ Microbiol 71(6):3033–3040

93. Chayabutra C, Wu J, Ju LK (2001) Rhamnolipid production by Pseudomonas aeruginosa under denitrification: effects of limiting nutrients and carbon substrates. Biotechnol Bioeng 72 (1):25–33

94. Davis DA, Lynch HC, Varley J (1999) The production of surfactin in batch culture by Bacillus subtilis ATCC 21332 is strongly influenced by the conditions of nitrogen metabolism. Enzym Microb Technol 25(3–5):322–329

95. Wei YH, Wang LF, Changy JS, Kung SS (2003) Identification of induced acidification in iron-enriched cultures of Bacillus subtilis during biosurfactant fermentation. J Biosci Bioeng 96 (2):174–178

96. Abouseoud M, Maachi R, Amrane A, Boudergua S, Nabi A (2008) Evaluation of different carbon and nitrogen sources in production of biosurfactant by Pseudomonas fluorescens. Desalination 223(1–3):143–151

97. Kim HS, Jeon JW, Kim SB, Oh HM, Kwon TJ, Yoon BD (2002) Surface and physico-chemical properties of a glycolipid biosurfactant, mannosylerythritol lipid, from Candidaantarctica. Biotechnol Lett 24(19):1637–1641

98. Makkar RS, Cameotra SS (1998) Production of biosurfactant at mesophilic and thermophilic conditions by a strain of Bacillus subtilis. J Ind Microbiol Biotechnol 20(1):48–52

99. Cooper DG, Paddock DA (1984) Production of a biosurfactant from Torulopsis bombicola. Appl Environ Microbiol 47(1):173–176

100. Sen R, Swaminathan T (1997) Application of response-surface methodology to evaluate the optimum environmental conditions for the enhanced production of surfactin. Appl Microbiol Biotechnol 47(4):358–363

101. Mukherjee S, Das P, Sivapathasekaran C, Sen R (2008) Enhanced production of biosurfactant by a marine bacterium on statistical screening of nutritional parameters. Biochem Eng J 42 (3):254–260

102. Wei YH, Lai CC, Chang JS (2007) Using Taguchi experimental design methods to optimize trace element composition for enhanced surfactin production by Bacillus subtilis ATCC 21332. Process Biochem 42(1):40–45

103. Fontes GC, Finotelli PV, Rossi AM, Rocha-Leão MHZ (2012) Optimization of penicillin G microencapsulation with OSA starch by factorial design. Chem Eng Trans 27

104. Najafi AR, Rahimpour MR, Jahanmiri AH, Roostaazad R, Arabian D, Soleimani M, Jamshidnejad Z (2011) Interactive optimization of biosurfactant production by Paenibacillus alvei ARN63 isolated from an Iranian oil well. Colloids Surf B: Biointerfaces 82(1):33–39

105. Kiran GS, Thomas TA, Selvin J (2010) Production of a new glycolipid biosurfactant from marine Nocardiopsis lucentensis MSA04 in solid-state cultivation. Colloids Surf B: Biointerfaces 78(1):8–16

106. Sen R, Swaminathan T (2004) Response surface modeling and optimization to elucidate and analyze the effects of inoculum age and size on surfactin production. Biochem Eng J 21 (2):141–148

107. Kosaric N, Cairns WL, Gray NCC, Stechey D, Wood J (1984) The role of nitrogen in multiorganism strategies for biosurfactant production. J Am Oil Chem Soc 61(11):1735–1743
108. Tahzibi A, Kamal F, MAZAHERI AM (2004) Improved production of rhamnolipids by a Pseudomonas aeruginosa mutant. Biodegradation 18:115–121
109. Ochsner UA, Reiser J, Fiechter A, Witholt B (1995) Production of Pseudomonas aeruginosa Rhamnolipid biosurfactants in heterologous hosts. Appl Environ Microbiol 61(9):3503–3506
110. Ohno A, Ano T, Shoda M (1995) Production of a lipopeptide antibiotic, surfactin, by recombinant Bacillus subtilis in solid state fermentation. Biotechnol Bioeng 47(2):209–214
111. Koch AK, Käppeli O, Fiechter A, Reiser J (1991) Hydrocarbon assimilation and biosurfactant production in Pseudomonas aeruginosa mutants. J Bacteriol 173(13):4212–4219
112. Lin S-C, Lin K-G, Lo C-C, Lin Y-M (1998) Enhanced biosurfactant production by a Bacillus licheniformis mutant. Enzyme Microb Technol 23(3–4):267–273. https://doi.org/10.1016/S0141-0229(98)00049-0
113. Iqbal S, Khalid ZM, Malik KA (1995) Enhanced biodegradation and emulsification of crude oil and hyperproduction of biosurfactants by a gamma ray-induced mutant of Pseudomonas aeruginosa. Lett Appl Microbiol 21(3):176–179
114. Mulligan CN, Chow TYK, Gibbs BF (1989) Enhanced biosurfactant production by a mutant Bacillus subtilis strain. Appl Microbiol Biotechnol 31(5–6):486–489
115. Shabtai YOSEF, Gutnick DL (1986) Enhanced emulsan production in mutants of Acinetobacter calcoaceticus RAG-1 selected for resistance to cetyltrimethylammonium bromide. Appl Environ Microbiol 52(1):146–151
116. Symmank H, Franke P, Saenger W, Bernhard F (2002) Modification of biologically active peptides: production of a novel lipohexapeptide after engineering of Bacillus subtilis surfactin synthetase. Protein Eng 15(11):913–921
117. Yakimov MM, Giuliano L, Timmis KN, Golyshin PN (2000) Recombinant acylheptapeptide lichenysin: high level of production by Bacillus subtilis cells. J Mol Microbiol Biotechnol 2 (2):217–224
118. Koch AK, Reiser J, Käppeli O, Fiechter A (1988) Genetic construction of lactose-utilizing strains of Pseudomonas aeruginosa and their application in biosurfactant production. Bio/Technol 6(11):1335–1339
119. Satpute SK, Banat IM, Dhakephalkar PK, Banpurkar AG, Chopade BA (2010) Biosurfactants, bioemulsifiers and exopolysaccharides from marine microorganisms. Biotechnol Adv 28 (4):436–450
120. Sarachat T, Pornsunthorntawee O, Chavadej S, Rujiravanit R (2010) Purification and concentration of a rhamnolipid biosurfactant produced by Pseudomonas aeruginosa SP4 using foam fractionation. Bioresour Technol 101(1):324–330
121. Sivapathasekaran C, Sen R (2017) Origin, properties, production and purification of microbial surfactants as molecules with immense commercial potential. Tenside Surfactants Detergents 54(2):92–107
122. Helmy Q, Kardena E, Funamizu N, Wisjnuprapto (2011) Strategies toward commercial scale of biosurfactant production as potential substitute for it's chemically counterparts. Int J Biotechnol 12(1–2):66–86
123. Alcantara VA, Pajares IG, Simbahan JF, Edding SN (2014) Downstream recovery and purification of a bioemulsifier from Sacchromyces cerevisiae 2031. Phil Agric Sci 96:349–359
124. Rau U, Nguyen LA, Roeper H, Koch H, Lang S (2005) Fed-batch bioreactor production of mannosylerythritol lipids secreted by Pseudozyma aphidis. Appl Microbiol Biotechnol 68 (5):607–613
125. Alonso S, Martin PJ (2016) Impact of foaming on surfactin production by Bacillus subtilis: implications on the development of integrated in situ foam fractionation removal systems. Biochem Eng J 110:125–133
126. Beuker J, Steier A, Wittgens A, Rosenau F, Henkel M, Hausmann R (2016) Integrated foam fractionation for heterologous rhamnolipid production with recombinant pseudomonas putida in a bioreactor. AMB Express 6(1):11

127. Willenbacher J, Zwick M, Mohr T, Schmid F, Syldatk C, Hausmann R (2014) Evaluation of different Bacillus strains in respect of their ability to produce Surfactin in a model fermentation process with integrated foam fractionation. Appl Microbiol Biotechnol 98(23):9623–9632

128. Winterburn JB, Russell AB, Martin PJ (2011) Integrated recirculating foam fractionation for the continuous recovery of biosurfactant from fermenters. Biochem Eng J 54(2):132–139

129. Küpper B, Mause A, Halka L, Imhoff A, Nowacki C, Wichmann R (2013) Fermentative production of monorhamnolipids in pilot scale–challenges in scale-up. Chemie Ingenieur Technik 85(6):834–840

130. Coutte F, Lecouturier D, Leclère V, Béchet M, Jacques P, Dhulster P (2013) New integrated bioprocess for the continuous production, extraction and purification of lipopeptides produced by Bacillus subtilis in membrane bioreactor. Process Biochem 48(1):25–32

131. Dolman BM, Kaisermann C, Martin PJ, Winterburn JB (2017) Integrated sophorolipid production and gravity separation. Process Biochem 54:162–171

132. Lemmens L, Colle I, Van Buggenhout S, Palmero P, Van Loey A, Hendrickx M (2014) Carotenoid bioaccessibility in fruit-and vegetable-based food products as affected by product (micro) structural characteristics and the presence of lipids: a review. Trends Food Sci Technol 38(2):125–135

133. Van Renterghem L, Roelants SL, Baccile N, Uyttersprot K, Taelman MC, Everaert B et al (2018) From lab to market: an integrated bioprocess design approach for new-to-nature biosurfactants produced by Starmerella bombicola. Biotechnol Bioeng 115(5):1195–1206

134. Roelants SLKW, Van Renterghem L, Maes K, Everaert B, Redant E, Vanlerberghe B et al (2019) Microbial biosurfactants: from lab to market. In: Microbial Biosurfactants and their Environmental and Industrial Applications. CRC Press, Boca Raton, pp 341–363

135. Henkel M, Müller MM, Kügler JH, Lovaglio RB, Contiero J, Syldatk C, Hausmann R (2012) Rhamnolipids as biosurfactants from renewable resources: concepts for next-generation rhamnolipid production. Process Biochem 47(8):1207–1219

136. Sekhon Randhawa KK, Rahman PK (2014) Rhamnolipid biosurfactants—past, present, and future scenario of global market. Front Microbiol 5:454

137. Sekhon KK, Khanna S, Cameotra SS (2012) Biosurfactant production and potential correlation with esterase activity. J Pet Environ Biotechnol 3(133):10–4172

138. Ruggeri C, Franzetti A, Bestetti G, Caredda P, La Colla P, Pintus M et al (2009) Isolation and screening of surface active compound-producing bacteria on renewable substrates. In: Current research topics in applied microbiology and microbial biotechnology. World Scientific, Singapore, pp 686–690

139. Saisa-Ard K, Manerrat S, Saimmai A (2013) Isolation and characterization of biosurfactants-producing bacteria isolated from palm oil industry and evaluation for biosurfactants production using low-cost substratess. BioTechnologia 94(3):275–284

140. Partovi M, Lotfabad TB, Roostaazad R, Bahmaei M, Tayyebi S (2013) Management of soybean oil refinery wastes through recycling them for producing biosurfactant using Pseudomonas aeruginosa MR01. World J Microbiol Biotechnol 29(6):1039–1047

141. Al-Bahry SN, Al-Wahaibi YM, Elshafie AE, Al-Bemani AS, Joshi SJ, Al-Makhmari HS, Al-Sulaimani HS (2013) Biosurfactant production by Bacillus subtilis B20 using date molasses and its possible application in enhanced oil recovery. Int Biodeterior Biodegradation 81:141–146

142. Sekhon KK, Khanna S, Cameotra SS (2011) Enhanced biosurfactant production through cloning of three genes and role of esterase in biosurfactant release. Microb Cell Factories 10 (1):49

143. Albarelli JQ, Santos DT, Holanda MR (2011) Energetic and economic evaluation of waste glycerol cogeneration in Brazil. Braz J Chem Eng 28(4):691–698

Industrial Applications of Biosurfactants

4

Abstract

Surfactants such as lecithin and egg yolk have been known for decades for their usefulness in the food systems. High toxicity and environmental influence of various surfactants have led to attention in other, natural surface-active agents such as microbial surfactants. The rising demand for surfactants in the food system also upsurges the consumer expectancy of non-toxic "green label" molecules over chemical surfactants. Sophorolipids and other established biosurfactants from Generally Regarded as Safe organisms have huge potential in the food system with an advantage of no acute toxicity. The major bottleneck is not the establishment of the cost-effective process but the apprehension about their safety and possible microbial hazards. Most of the research about biosurfactants in the food system are only limited to the applications, without doing an assessment of safety and risk analysis, and it is one of the undermined reasons behind the adoption of biosurfactants for food application. It should be consider for the appropriate applications such as biosurfactants with high HLB value must be used as food emulsifiers whereas, antimicrobial potential can be achieved at low concentrations and synergies with other conventional food preservatives. The efficacy of surfactin, rhamnolipids, and sophorolipids is evidenced to be utilized in bakery products with improved texture, properties, and food quality. The present book chapter emphases the most promising properties of biosurfactants, current use in food formulations, risk, and toxicity assessment, and forthcoming biosurfactants to substitute the currently used surfactants will be assessed.

Keywords

Food additives · GRAS · Antimicrobial · Acute toxicity · Biosurfactants

© Springer Nature Singapore Pte Ltd. 2021
D. Sharma, *Biosurfactants: Greener Surface Active Agents for Sustainable Future*,
https://doi.org/10.1007/978-981-16-2705-7_4

4.1　Introduction

Surfactants are categorized as the most multipurpose molecules explored in diverse applications in detergents, pesticides, food processing, pharmaceuticals, cosmetics, and petroleum recovery [1]. Surfactants are obtained from petrochemicals, animal fats, plants, and fatty acid esters [2, 3]. Naturally known surfactants have been used in the food systems such as lecithin and milk proteins in the production of mayonnaise, salad dressing, and deserts [4, 5]. The rising demand for surfactants, the characterization of molecules with low toxicity, and efficient surface activity potential are of extensive interest.

Chemical surfactants are linked with many drawbacks, out of that intestinal dysfunction is majorly reported. Naturally, various surfactants in high concentration are present in the foods which leads to the increased intestinal permeability and can elicit allergic and inflammatory responses [6]. There is no established acceptable daily intake (ADI) for surfactant in the food system [7]. Demands for green constituents over chemical additives ("green label") have led to wide research on screening new sources such as microbial surfactants.

Biosurfactants have conferred different functional attributes to food (emulsifying, additive, foam formation, wetting agents, antibiofilm) and functional properties (antimicrobial). Even so, biosurfactants have an incontestable potential for substituting the chemical surfactants with huge significance to the food system. The major bottleneck is not the establishment of the cost-effective process but the apprehension about their safety. Most of the research about biosurfactants in the food system is only limited to the applications, without assessing safety and risk analysis, and it is one of the undermined reasons behind the adoption of biosurfactants for food application [8].

Furthermore, the regulation governing new additives requires extensive research and investigation to secure approvals for inclusions. Dedicated guidelines to adopt biosurfactants in the food system do not exist, but the guidelines to include general food additives might be adopted to charter the primary requirements. In the present book chapter, we have discussed the potential roles of biosurfactants as bio emulsifiers and food additives, major challenges, forthcomings considering regulatory qualifications.

4.2　Biosurfactants Role in Food

The food sector industry can make use of microbial surfactants in two different ways: (a) indirectly they can be utilized for the pre-conditioning or cleaning of food contact surfaces and (b) directly they can be included as food additives or ingredients. On the other hand, the food processing industry can make use of their generated wastes which are generated as potential feedstocks for the production of biosurfactants, thus contributing towards the valorization of waste and therefore resulting in the reduction of costs required for waste treatment [9, 10]. In recent times, an increasing number of reports have been published regarding the use of biosurfactants for the control of food pathogens, and the biosurfactant exploitation as an additive has gained enormous interest.

4.2.1 Properties of Biosurfactants Ideal for the Food Applications

- Antimicrobial properties
- Antibiofilm/food sanitations
- Food additives
- Food emulsifications
- Food preservation
- Antioxidants

4.2.2 Surface Pre-conditioning by Microbial Surfactants

Biofilm can be well-defined as a heterogeneous group of microbial communities that are embeddesd within the extracellular polymeric matrix (EPM) and are attached to the biotic or abiotic surface. Biofilms protect microorganisms from various hostile environments. The biofilm cells possess higher resisting power towards physical and chemical compounds as compared to that of planktonic cells because biofilm cells act as a blockade that either averts or reduces contact with antimicrobials agents [11]. The onset of biofilm takes place progressively, dynamic, and successively, and consists of adhesion, maturation, and dispersion [12]. The initial step includes the closeness of planktonic microorganisms with that of a surface, therefore the physical and chemical characteristics of cells and surface are the two main factors for the rate and level of adhesion course. Additionally, chemical and physical modification can take place due to the formation of the conditioning layer, which acts as a molecule coat thereby promoting or inhibiting microbial adhesion. This layer is formed by the exposure of the surface to that of an aqueous medium [13]. Reversible adhesion relies on the cell surface properties like the existence of appendages such as fimbriae and flagella, hydrophobic in nature, and associated proteins or sugar that might have an impact on the electrostatic interactions in nutrient and cell [14]. Irreversible surface adherence takes place when an adhered microorganism produces adhesins and exopolymers [15]. The subsequent phase is growth, micro-colonies production, and attachment to the local environment supplementary microbial cells [11]. In general, biofilms can develop on most surfaces which include plastic sheets, glass, metal surfaces, wood, and food products, etc. Food processing involves various equipment surfaces that provide favorable environments for the formation of biofilm, principally where both moisture and nutrients are easily accessible [16]. The onset of biofilms is of great concern in food processing environments because they lead to the spoilage of food products and hence put public health at risk; so, their occurrence is of great apprehension for the food processing industry.

Apart from generating several serious difficulties for the process actions and food spoilage, there are also higher chances of food-associated disease transmission due to the adherence of pathogenic bacteria like *Listeria monocytogenes, Escherichia coli*, and *Salmonella spp* to the surfaces [11].

Chemical sanitizers have shown their effectiveness against planktonic cells, but in the case of the removal of biofilms, they are less effective because the cells are embedded within the EPM which provides them with protection [14]. Hence, the reduction of biofilm onset is necessary to provide harmless and good quality food products to customers. Food processing can employ biosurfactants to prevent, inhibit, and remove biofilm-producing microorganisms.

4.2.3   Biosurfactants as Food Additives

As per the European Food Safety Authority (EFSA), food additives are defined as a substance which can be intentionally added to foodstuffs to bring some special functions. Food-based additives are considered as preservatives, nutritional additives, flavoring enhancers, coloring material, texture improver, and miscellaneous agents [17]. Antimicrobials and antioxidants are potential examples of food preservatives, while functional additives such as vitamins, dietary fibers, and essential amino acids are some classes of additives. Flavoring compounds consist of sweeteners, aroma, and flavor improvers, while texture improvers consist of emulsifiers and stabilizers, typically incorporated to enhance either food texture or mouthfeel of food preparations. Miscellaneous food additives consist of chelating compounds, lubricants, food modifying enzymes, and also anti-foaming substances [17]. In the middle of the twentieth century, additives were introduced and directly associated with food industrialization and formulations [18], and from that time, there is a considerable increase in the demand for food additives. According to [19] estimated that the food additives market is to cross about US\$39.85 billion by 2021. Synthetic additives are being replaced with novel bio-based molecules because of the growing awareness of consumers about synthetic products along with the increasing attention for natural, organic, plant origin, and other consumer-specific foods. Subsequently, biosurfactants are acceptable choices to satisfy the future market patterns, since they have exhibited valuable characteristics to be utilized in food formulations.

Biosurfactants as a food additive also include substances that may be introduced to food directly (emulsifiers, foaming, thickening, texture improvement, preservatives, encapsulation of fat-soluble vitamins) (called "direct additives") in the manufacturing process. Biological activity such as antiadhesive/antimicrobial or food surface sanitations (called "indirect additives") through packaging, coating, or during storage (Table 4.1). But, before the inclusion to food systems, it must undergo toxicological evaluation, considering synergies with food molecules, and protective effects stemming from its use. A report by Haesendonck and Vanzeveren [40] emphasized the improvement of dough texture and volume in bakery products after adding rhamnolipids. Rhamnolipid could be hydrolyzed commercially to

Table 4.1 Various reports demonstrated the possible role of the biosurfactants in the food system

Antiadhesive			
Microorganisms	Biosurfactant	Pathogens	Reference
Pseudomonas putida	Putisolvin I and II	*Pseudomonas* sp.	Kuiper et al. [20]
	Pseudobactin II	*Enterobacter faecalis, Proteus mirabilis*	Janek et al. [21]
Bacillus subtilis	Fengycin	*Salmonella enterica*	Rivardo et al. [22]
Bacillus tequilensis	Lipopeptide	*Streptococcus mutans*	Pradhan et al. [23]
Candida sphaerica	Lunasan	*Streptococcus agalactiae*	Luna et al. [24]
Pseudomonas aeruginosa	Rhamnolipid	*Yarrowia* sp.	Dusane et al. [25]
Candida lipolytica	Rufisan	*Streptococcus* sp.	Rufino et al. [26, 27]
Serratia marcescens	Glycolipid	*Candida albicans, Pseudomonas aeruginosa*	Dusane et al. [28]
Antimicrobial			
Microorganisms	Biosurfactant (MIC µg/mL)	Pathogens	Reference
Pseudomonas aeruginosa	Rhamnolipids (4–64)	*Alternaria alternata Aureobasidium pullulans Aspergillus Niger Chaetomium globosum Gliocladium virens*	Benincasa et al. [29]
Pseudomonas aeruginosa	Rhamnolipids (20–50)	*Alternaria mali Botrytis cinerea Fusarium sp. Rhizoctonia solani*	Kim et al. [30]
Pseudomonas aeruginosa	Rhamnolipids (0.5–1.70)	*Botrytis cinerea Fusarium sp. Gliocladium virens Penicillium funiculosum Rhizoctonia solani*	Haba et al. [31]
Pseudomonas aeruginosa	Rhamnolipids	*Botrytis cinerea*	Varnier et al. [32]
Pseudomonas aeruginosa	Rhamnolipids (64–256)	*Mucor miehei*	Nitschke et al. [33]
Pseudomonas sp.	Rhamnolipid	Pseudomonas aeruginosa	Sotirova et al. [34]
Food additives			
Microorganisms	Biosurfactant	Applications	Reference
Bacillus subtilis	Surfactants	Emulsifier	Chander et al. [35]
Candida utilis	–	Mayonnaise emulsification	Campos et al. [36]
Bacillus subtilis	–	Bakery additive	Zouari et al. [37–39]
Bacillus subtilis	Surfactin	Food preservation	Sharma et al. [5]
Pseudomonas sp.	Rhamnolipid	Dough improvement	Haesendonckand Vanzeveren [40]

(continued)

Table 4.1 (continued)

Antiadhesive			
Microorganisms	Biosurfactant	Pathogens	Reference
Pseudomonas aeruginosa	Rhamnolipid	Flavor enhancer	Trummler [41]
Nesterenkonia sp.	Lipopeptides	Texture improvement	Kiran et al. [42]
Bacillus subtilis	–	Cookies dough	Zouari et al. [37]
Bacillus subtilis	Lipopeptides	Bread improver	Mnif et al. [43]
Candida bombicola	Glycolipids	Cupcakes additive	Silva et al. [44]
Starmerella bombicola	Sophorolipids	Sophorolipids + curcumin	Vasudevan and Prabhune [45]
Probiotic (GRAS)	–	Animal fodder	Konkol et al. [46]
Emulsification			
Microorganisms	Biosurfactant type	Emulsification material	Reference
Bacillus vallismortis	Exopolysaccharides	Essential oils	Song et al. [47]
Nesterenkonia sp.	Lipopeptide	Unsaturated hydrocarbons	Kiran et al. [42]
Candida utilis	Glycolipids	Vegetable oil	Campos et al. [48]
Pseudomonas aeruginosa	Rhamnolipids	Saturated hydrocarbons	Nitschke et al. [49]
Kluyveromyces marxianus	Mannoprotein	Corn oil	Nitschke and Costa [50]
Saccharomyces lipolytica	–	Cooking vegetable oil	Lima and Alegre [51]
Pseudomonas aeruginosa	Rhamnolipids	Nano-emulsion	[52]

produce L-rhamnose, as a high-quality flavor compound. L-rhamnose is a methyl pentose natural sugar and is found in rhamnolipid [41].

Kiran et al. [42] reported the lipopeptide of *Nesterenkonia* sp. for the improvement of muffin texture. The muffin showed a reduction in hardness, chewiness, and gumminess as compared to control, at about, 0.75% lipopeptide concentration.

Campos et al. [36] demonstrated the production of mayonnaise with biosurfactants of *Candida utilis* as a major ingredient, and bestowed stability to the emulsion during storage. Dough texture and volume were significantly improved as compared to glycerol monostearate at the concentration of 0.1% biosurfactants. *Bacillus subtilis* surfactants conferred dough improvement and inclusive quality of cookies [37]. Similarly, Mnif et al. [43] also reported the improvement of dough quality on the addition of 0.075% (w/w) lipopeptides on the quality of bread in comparison to soya lecithin. Certainly, it led to an improvement in dough volume with improved texture profile of bread, chewiness, cohesion, and a decrease in firmness.

In another utilization of biosurfactants in a bakery, Silva et al. [44] demonstrated the production of cupcakes containing biosurfactants. The biosurfactant was incorporated into cupcakes, substituting 50%, 75%, and 100% of the plant fat. The substitution of plant fat by biosurfactant can also lead to improvement of nutritional value by decreasing the trans fatty acids content in plant fat. In 1951, it was established that surfactants encourage the growth of chickens [53]. The biotransformation of rapeseed meal using GRAS microorganism improved nutritional parameters and it was found enriched with biosurfactants. Biosurfactants hydrolyzed rapeseed meal has probiotic benefits and can be used as a substitute for antibiotics, which are prohibited from animal feed [46].

4.2.4 Biosurfactants as Food Emulsifying Agents

Various food systems have broadly employed surfactant molecules as emulsifiers. During the process of emulsification, some of the low-mass surfactants are amphiphilic molecules like monoglycerides, lecithins, glycolipids kind of surfactants, fatty alcohols, and fatty acids and can reduce interfacial tension efficiently. Proteins such as gelatin, milk casein, whey-based proteins, and sugars as alginate, gum arabic, pectin, xanthan gum, dextran, starch, and plant cellulose derivatives are natural surfactants and are intricated in food emulsion developments and stabilization, though, plant-based surfactants do not display effective surface activity [54, 55]. Food emulsion preparations play a critical role in the solubilization and dispersion of fats and also help in getting steadiness, textural appearance, and structure [56].

The primary function of an emulsifier is to stabilize an emulsion by controlling phase agglomeration. Surfactants (or biosurfactants) can enhance this stability by reducing the interfacial tension, which in turn results in reducing the surface energy between the two phases [50]. According to the characteristics stated earlier, in comparison to synthetics, biosurfactants possess a great role in food manufacturing. Microbial biosurfactants have a significant effect on important functions like stabilization of aerated systems, helps in improving the consistency, extension of product's shelf life, and improves the rheological characteristics; however, only the utilization of microbial surfactants as food additives is previously explored in few studies.

Some patents have been initially granted for the utilization of rhamnolipids in food preparations to improve the shelf life of bakery and dairy foods', dough steadiness, volume as well as stability [57, 58]. In previous years, Mnif et al. [43] studied the effect of bread quality on the addition of lipopeptide biosurfactant. It was observed that the addition of a concentration of about 0.075% biosurfactant emulsifier has pointedly enhanced specific volume and crumb pattern & structure in comparison to the same formulations utilizing soy lecithin. The BSs also aid in enhancing the texture, reduced staling, and also on storage of 8 days microbial proliferation was seen decreased. The incorporation of *Bacillus subtilis* derived surfactin as an emulsifier in cookies formulation has also been documented. The biosurfactant emulsifier concentration of about 0.1% showed better properties regarding the dough texture like hardness, stickiness, and springiness on comparing with the dough texture added with a synthetic emulsifier such as glycerol monostearate [37–39].

Suresh Chander et al. [59] have established that *B. subtilis* strain-derived biosurfactant was capable of emulsifying various edible oils, therefore presenting the great ability to be used in food systems. The polymeric surfactant structure including polysaccharide, fatty acid, and protein fractions are also assigned as bio emulsifiers because they provide well emulsifying capacity and the potential to stabilize emulsions [60]. Bioemulsifiers derived from yeasts, particularly from genera *Candida, Pseudozyma,* and *Yarrowia,* are generally recognized as safe (GRAS), which is regarded as a vital standard for food application as related to bacterial BSs [48]. A yeast strain, i.e. *Candida utilis* has been documented to deliver BSs with a potential to be utilized as a bio emulsifier in different food preparations [48]. In another study, *C. utilis* derived biosurfactant was used to formulate mayonnaise based on sunflower oil. Mayonnaise product obtained after incorporated with biosurfactant and guar gum showed stability for about 30 days at 4 °C.

To determine whether *C. utilis* derived biosurfactant is safe for utilization in food processing an in vitro toxicity test was done. The test was done on rats by feeding them with biosurfactants and results showed no toxic effects on rats. Thus *C. utilis* derived biosurfactant has the potential to be used in food processing [36]. Biosurfactants can be used in the preparations of nano and microemulsions which act as a carrier of various food constituents including vitamins, probiotics cells, and functional metabolites [61]. In food systems, nano-emulsions can have other applications as well such as cleaning of equipment surfaces or as food preservatives. Antimicrobial activity was shown by surfactin– sunflower oil nano-emulsion against various food-associated pathogens like *Staphylococcus aureus, Listeria monocytogenes, Salmonella typhi*, and *Bacillus cereus* spores. The microbial population was significantly reduced when surfactin-based nano-emulsion was utilized in vegetables, meat products, the milk industry, and apple juice [62]. Recently, Farheen et al. [63] proposed nano-biosurfactant emulsion involving *P. aeruginosa* derived biosurfactant. According to the author, nano-biosurfactant preparations had shown significant emulsifying activity than that of chemical surfactant when explored to butter.

Emulsification capacity is also predicted by understanding its hydrophilic-lipophilic balance (HLB), which designates biosurfactant potential to form water-in-oil (w/o) or oil-in-water (o/w) emulsion. The HLB has been judged on a scale of 0–20. Biosurfactants for appropriate applications can be categorized on HLB values. Emulsifiers with high HLB confers better solubility of oil-in-water. Generally, HLB values ranges (3–6) indulge in w/o microemulsions, while HLB values (8–18) encourages o/w microemulsions. For example, surfactin and rhamnolipids, as their HLB range favor the development of o/w emulsions (Table 4.2) [66].

Table 4.2 HLB values comparison of different surfactants

Type of biosurfactants	HLB value	Reference
Rhamnolipids	10.17	Khoshdast and Sam [64]
Sophorolipids	10–13	Vaughn et al. [65]
Surfactin	10–12	Gudiña et al. [66]
Mannosylerythritol lipids	≥12	Randu et al. [67]
Other glycolipids	10–15	Sekhar and Nayak [68]
Lipopeptides	10–11.1	De Zoysa et al. [69]
Polysorbate 80	14.4–15.6	Braun et al. [70]

The key potential of surfactants is reducing the interfacial tension, which permits the establishment of small droplets. The reduction in emulsion droplet size, which leads to improved stability [71]. Another potential application is microemulsions which can be a carrier for fat-soluble vitamins, and value-added molecules [61]. Nano-biosurfactants based on rhamnolipids have been reported by Farheen et al. [63], with improved emulsifying potential as compared to the chemical surfactants.

Currently, enormous knowledge is available regarding the use of common chemical surfactants for the preparation of drug delivery systems; therefore, replacement of chemical surfactant with somewhat similar molecules but the microbial origin is anticipated to be an easy and simple approach as well as to offer novel prospects given to the presently used chemical molecules [72]. Among the diverse nano-sized delivery systems, micro-emulsion-based colloidal ones can be developed with the help of biosurfactants. The main components required for the development of those systems involve an aqueous phase, an oil phase, a surfactant, and at times a co-surfactant or co-solvent. The biosurfactants can self-assemble into different forms or structures that are capable to encapsulate and thereby solubilize the hydrophobic or hydrophilic compounds in the vicinity of the dispersed phase (oil for O/W and water for W/O microemulsions) inside its structural core, therefore separating the dispersed phase from that of the continuous phase [73]. Mannosylerythritol lipids have shown a higher rate of emulsification with soybean oil and tetradecane than that of polysorbate 80 [74]. Moreover, the formation of stable W/O microemulsion without the addition of any co-surfactant or salt is another remarkable property of these biosurfactants [75]. Also, other BSs, such as rhamnolipids and sophorolipids can be mixed individually with lecithins to get biocompatible microemulsions [76]. As Liu et al. [77] studied the application of rhamnolipids in the development of emulsion-based fish oil delivery systems which contain high levels of ω-3 polyunsaturated fatty acids for fusion with foods. Rhamnolipids displayed better results concerning the protection of polyunsaturated fatty acid from oxidation than that of natural emulsifiers such as saponins. Microbial surfactants have been proposed to be utilized for the preparation of nanoparticles and liposomes apart from the formulation of microemulsions [78]. In recent times, interest in biosurfactants has increased mainly because of their potentiality to be used in the preparation of silver nanoparticles and NiO nanorods [79, 80]. Reddy et al. [81] have stabilized and synthesized the silver nanoparticles with surfactin. Rhamnolipids have been also examined for their potential role in the synthesis and stabilization of nano-zirconia particles [82]. BSs are considered safe, multipurpose, and useful for a large number of applications which even includes the production of nano-solutions for the food industry. Biosurfactants have shown high biocompatibility as well as low toxicity. Rhamnolipids, surfactin, iturin, and pumilacidin are some of the biosurfactants which are employed for the preparation of oral lipid-based preparation of therapeutic substances [66]. To enhance the biological performance of vitamin E biosurfactant surfactin has been used for its self-micro emulsifying drug delivery system. This delivery system has shown a surprising increase in the emulsification activity, rate of dissociation, and ultimately oral bioavailability [72]. Moreover, siRNA delivery in HeLa cells was enhanced with the help of cationic surfactin [83]; mannosylerythritol lipid vesicles have shown high levels of gene transfection efficacy [84–86]; and rhamnolipids were utilized for the preparation of drug encapsulated lipid-polymer coated hybrid nanoparticles for controlling the biofilm formation [87]. Though, regardless of all the advantages of utilizing BSs for such purposes, the huge potential remains unexplored, in specific for the food industry.

Sophorolipids are known for their significant emulsification of vegetable and oil used in bakery formulations. Gaur et al. [88] reported sophorolipids by Candida and their applications as a food emulsifier. Biosurfactants showed significant emulsification with olive oil (51%), soybean oil (39%), almond (50%), and mustard (50%). It was established that biosurfactants act as an emulsifying molecule, which also efficiently emulsifies vegetable oils. Various other "indirect" applications such as antimicrobial, food preservation, and antiadhesive activity need more validation and standardization of the methods, environmental factors, and other synergies [5].

4.2.5   Biosurfactants Role in Food Preservation

Microbial surfactants are considered and recognized as efficient antimicrobial agents such as glycolipids, surfactin, lipopeptides, fengycin, iturin, and bacillomicins, and having a huge potential to control foodborne pathogens. But there are certain issues like hemolytic behavior of surfactin [89]. BSs derived from the various opportunistic pathogens like *Pseudomonas aeruginosa* are restricted to be utilized in food processing and henceforth direct application of the BSs in food formulation is restricted. All the same, certain efforts have been made regarding the usage of BSs to eliminate food pathogens in different environments. The population of the pathogens like *Salmonella enteritidis* was significantly decreased in milk samples with Surfactin and polylysine synergistically in a ratio of 1:1 [90].

A certain patent has been filed for rhamnolipids utilization in the cleaning of vegetables and fruits [91]. Surfactants obtained from the Lactobacillus and food-associated yeasts may epitomize a safer substitute for food processing and formulations due to lower toxicity. Sambanthamoorthy et al. [92] reported the BSs derived from *Lactobacillus jensenii* and *Lactobacillus rhamnosus* were found potential to control the growth of *Escherichia coli* and methicillin-resistant *Staphylococcus aureus*. BSs obtained from the GRAS yeast like *Candida apicola* and *Candida bombicola*, displayed lower toxicity as compared to the surfactin obtained from the Bacillus genus [93]. Various reports confirmed the role of glycolipids to control against various microorganisms and can be explored for vegetable and fruit sanitization ([5, 94]).

The sophorolipids derived from the yeast *Wickerhamiella domercqiae* was reported for fruit preservation at ambient temperature [95]. Increase of membrane permeability, creation of porins, membrane solubilization, and ionic adducts are some major mechanisms of BSs mediated antimicrobial potential [96–98]. Planktonic cells of a pathogenic strain of *L. monocytogenes* were inhibiting by rhamnolipids [99]. Rhamnolipids displayed significant antimicrobial activity at a concentration of 78.1–2500 mg/mL [100]. In the present case, BS and bacteriocins as a synergistic combination demonstrated significant antimicrobial potential. The synergism displayed here is due to both microbial metabolites worked on the plasma membrane. Bacteriocins are also recognized for their role in food preservation and one of the most explored food preservatives. Such results epitomize new potentials on applications of BSs such as their synergy with bacteriocins can enhance the production of active preservatives for food packaging, and discovering their antiadhesive, antibacterial, antifungal, and synergistic effects.

4.2.6 Microbial Surfactants in Food Sanitation

Pathogenic microorganisms are predominantly survived on food processing and handling surfaces, food handling, and processing types of equipment and vessels [101, 102]. There are various strategies utilized for the removal of the pathogenic populations found on the food surfaces [103]. The incidence of pathogenic biofilms development on food surfaces may consequently lead to ensuing events:

• Dispersal of foodborne pathogens
• Food spoilage
• Contamination of the food by non-indigenous starter
• Oxidization of metal in contact with the food such as supply and transport and storage tanks

The contamination and biofilm formation by the pathogenic bacteria indicate a vital problem in the food processing surfaces [8, 102]. Different food preparation and handling practices may be responsible for the low level of food safety and hygiene where microbial biofilms can colonize ultrafiltration equipment, filtration membranes, hollow fibers, metal surfaces, food-grade silicone seals, and tubing [104]. Food storage and biofilm formation are sometimes directly parallel to each other such as the presence of various foodborne pathogens, *L. monocytogenes*, *L. innocua*, *Salmonella* spp., and *Staphylococcus aureus*, and food intoxication by *Bacillus cereus* are huge apprehension in food preparations and processing

[5, 105]. In food formulation, BSs offer different properties and play a role in emulsifying reactions, antibacterial, antioxidant agents, and antibiofilm activities. The utilization of the BSs is a futuristic approach for combating established biofilms is one of the key areas of challenge in the food processing sector [1, 99, 106]. Ideally, BSs were explored for hydrocarbon uptake and bioremediation.

Though BSs have been recognized as a substitute for chemical-based surfactants for the prevention of pathogenic biofilms, specifically in food-formulations and sanitation [50]. The existing approaches such as physical and chemical have different drawbacks and are even risky to food safety in certain cases. The control of bacterial biofilms establishment could be achieved using various constituents or alteration of surfaces.

Microbial surfactants can even adapt and prevent the biofilm on a metal surface such as stainless steel [107]. In addition to the pre-coating of food, surfaces can also be utilized in the prevention of biofilm on food processing surfaces [105, 108]. BSs derived from the *Pseudomonas fluorescens* diminished the biofilm of *L. monocytogenes* on a stainless steel surface. The establishment of Listeria sp. was allowed on the stainless steel surface and further exposed to the BSs derived from the Lactobacillus *helveticus* [109].

Along with the BSs obtained from the probiotic strains, surfactants derived from the surfactin and rhamnolipids have been utilized for the control of the biofilms. BS obtained from the rhamnolipid and surfactin producer strains are being observed for their probable antibiofilm activity against various food contact pathogens on poly-propylene and metal surfaces [110]. Pre-conditioning of the food surfaces with the BSs decreased the contact angle and also results in the low degree of cell appendages responsible for the anchored pathogenic biofilms. Additionally to the antimicrobial potential of probiotic BSs, also contributes to the biofilm prevention to control various foodborne pathogens such as pathogenic yeast and human pathogens [111, 112]. An additional explanation is that BSs dissolves the exopolysacchrides secreted by the biofilm-forming pathogens.

Different BSs obtained from various microbial pathogens were used to prevent the biofilm development on food and biomedical surfaces and significantly reduces the chances to biofilm development. BSs displayed potential usage in cleaning different food processing like stainless steel, polystyrene materials, heat exchangers plates, and transport pipelines. Such potential of the BSs biosurfactants displayed and recommended its appropriate usage in dairy, meat processing, and other food formulating sectors. Usually, the BSs are considered as non-toxic, but still, BSs toxicity needs to be resolute for their effective applications in food preparation and formulations.

4.3   Sensorial Behavior

Organoleptic assessment of biosurfactants is a key parameter of food additives to be qualified for inclusion in food systems. Oral acceptance is indicative of acceptance of biosurfactants as a food additive, especially on taste and mouthfeel, and it may lead to lower conviction among a large majority of consumers [113]. The information about the sensorial acceptance of biosurfactants is still in its infancy. Ozdener et al. [114] demonstrated a sophorolipid biosurfactant obtained from *S. bombicola* for sensorial properties. It is also completely possible that biosurfactants produced by oral microbiome or GRAS microorganisms available from food sources. The results showed possibilities for applied applications for biosurfactants to improve the bitter perceptions of foods and improve acceptance by consumers. The combined properties of biosurfactants and sensory taste stimulants will significantly increase the commercial value of biosurfactant molecules. Sensorial acceptance will open new promises for applied applications concealing/blocking agents for the bitter sensitivity of foods. Bitter-taste discernment is innate and encourages aversive reactions and reducing the bitter-taste discernment is vital for the acceptability of food. In another report, antifungal biosurfactant produced by Bacillus sp. as starter culture during fermentation of Maari beverage of west Africa was found satisfactory for sensorial properties [115]. Scientific and sensorial attributes are strain-dependent, and it may help to differentiate strains with technological properties and performances.

4.4   Food Matrix Interactions

Despite the huge potential of the biosurfactants in food industries, it would be dishonest not to discuss the efficiency of the biosurfactants without debating their interaction/synergies with food molecules. High molecular weight (HMW) biosurfactants cover protein and polysaccharide groups present in any food system. Headgroups of the molecule bind to groups of contrary charge on the protein. HMW biosurfactants are efficient emulsifying agents. The anionic surfactant binding is pH-regulated and appears to be measured by the cooperative hydrophobic exchanges.

Some of the ideal characteristics of bio emulsifiers for industrial use are: (1) lowest amount vital to produce small droplets; (2) lowest droplet size; and (3) stability at different pH, ionic strength, and temperature [116]. For attaining the antimicrobial potential of the biosurfactants, their synergistic potential would be determined as most of the non-specific antimicrobials acts on the cell membrane. Magalhães and Nitschke [100] demonstrated the synergistic bacteriostatic activity of biosurfactants when conjugating with the nisin. The authors recommended that the synergism detected could be since both molecules acting on the cell membrane. Most of the fermented food is inhabited by lactobacilli and biosurfactants producing *Bacillus subtilis* such as natto and plant-based fermentation [117]. The selection of the right matrix, low doses, and maximum daily intake not only increases the efficacy but also contributes to overcoming the issues to their chemical counterparts. The combination with bacteriocins and other conventional food preservatives can progress the production of active antimicrobial food packaging, antimicrobial, and synergistic effect. Moreover, sophorolipids are anticipated to form self-assemblies with exceptional functionality. This directs the assumption that glycolipids being amphiphilic in nature can permeate to the structurally similar cell membrane and enable the entry of low concentrations of antimicrobials to achieve the desired potential without causing substance abuse.

4.5   Regulations to Commissioned Biosurfactants as a Food Additive

The selection of an emulsifier is essential to be considered as food additives, the biosurfactants essentially hold functional properties to achieve droplet coalescence and they must be non-toxic [118]. There are no regulatory guidelines known to charter any biosurfactants as a food additive. To minimize the time to commercialize any research outcome to pre-commercialization stage, we must stick to the regulatory framework used for a general food additive. Acute toxicity (LD_{50} dose) and allergy tests are needed to be qualified as a food additive [54]. The significance of the major endpoints as mentioned by the Organization for Economic Co-operation and Development (OECD) rules for the risk valuation of food chemicals is censoriously analyzed based on acute toxicity, allergic reactions, reproductive toxicity, and mutagenicity analysis can be adopted for the biosurfactants safety assessment.

The addition of 0.7% (w/v) glycolipids in mayonnaise production have been demonstrated. The administration of approximately 15 g of food by an adult of 50 kg weight would be equal to 0.10 g of surfactant, which would agree to 2 mg/kg of total body weight. The present dose of 2 mg/kg is much lower than that administered to determine acute toxicity in tested animals (3600 mg/kg), which led to conclude no acute risk [36]. US Food and drug administration (FDA) and Agriculture Organization and the World Health Organization directed that an acceptable daily intake (ADI) should be recognized that would deliver "an adequate margin of safety to reduce to a minimum any hazard to health in all groups of consumers." The ADI approximates the amount of a food additive articulated based on body weight that can be consumed daily over a period with no considerable health risk. ADIs are only assigned to those food additives that are considered clean from the body within 24 h.

A xylolipid produced by *Lactococcus lactis* also analyzed for acute toxicity as directed by the Office of Pollution Prevention and Toxics 870.1100; 152–10 [119]. No evident and quantifiable sickness or death occurred throughout the study phase. So, the LD_{50} of the isolated BSs was >5000 mg/kg of animal weight. Henceforth, biosurfactants were put in the toxicity class IV of FDA, and are regarded as safe for oral usage.

Industrial production of rhamnolipids as a food additive has started in the USA (Jeneil Biotech, USA), with no described hazards. Bio-surfactants, obtained from *Candida starmella*, GRAS approved strain and its industrial production of sophorolipids is also ongoing in Asian countries [120]. A Germany-based company Evonik has started the production of sophorolipids on an industrial scale. According to Evonik, sophorolipids complied the necessities of OECD 301 F (aerobic biodegradability of organic chemical) and ISO 11,734 (anaerobic biodegradability) with Renewable Carbon Index (RCI) of 100% and achieves suggestively better than chemical surfactants when evaluated for the water toxicity (OECD 211 and 202). Sophorolipids production with enhanced yields using genetically manipulated yeast and decrease the production costs was efficaciously granted a US Patent.

Food constituents may be additives after getting approval from the USFDA for explicit uses or when having the label of "GRAS." Biosurfactants producing *Candida utilis* is listed as GRAS organisms as per the Code of Federal Regulations, which originates somewhat from FDA regulations, under 21 (21CFR-172.590), which comprises permitted food additives. Furthermore, microorganisms originated constituents may be the focus of a GRAS notice [121]. *Candida starmella, Bacillus subtilis,* and lactic acid bacteria could be used to establish the food additives guidelines to streamline biosurfactants as they all are approved GRAS strains.

4.6   Role of Emulsions in Pharmacy and Cosmetic Industry

4.6.1   Pharmaceuticals

The development of nano-emulsion and emulsions are in huge demand for pharmaceuticals and the food industry. Various drug delivery systems are based on emulsion preparation such as transdermal, nasal, ocular, oral, and parenteral. Phospholipids are generally utilized appropriately as an effective emulsifier utilized in pharmaceutical preparations. Chemically synthesized and naturally occurring phospholipids are utilized as the constituent of drugs administrated as emulsions, liquid suspensions, or dispersion of solids. The synthetic phospholipids are utilized for injectable drugs whereas, natural phospholipids can be explored for various drug transport routes such as parenteral preparations.

Drug components or fat-soluble vitamins are specifically utilized for the production of surfactants-based preparation due to their limited water solubility. For example, Tween 20 and polypropylene glycol were utilized for the preparation of the o/w nano-emulsion of an anti-cancer drug, i.e. paclitaxel due to its limited solubility in water. Various antiviral drugs such as HIV or antiretroviral preparation has been initially developed as nano-emulsions.

4.6.2   Cosmetics Formulations

The role in the delivery of functional molecules in the cosmetics industry is critical and relied on the effectiveness of the emulsion/nano-emulsion. The appearance, texture, and softness of the preparation of the cosmetic are the vital parameters of the product [122]. The selection of the surface-active agents for the preparation of the cosmetic is anticipated to reduce the skin irritations, potential allergic response, interactions of the skin keratinocytes which result in inflammation. Various synthetic surfactants like non-ionic ethoxylated fatty acids and alcohols, alkyl polyglucosides, polyglycerol esters, poly-ethylene glycol (PEG), oligo(ethylene glycol); combinations of fatty alcohols, and sugar can be utilized in cosmetic formulations.

The utilization of the combinations or blends of surfactants is a general exercise that can offer cosmetics preparation with decreased irritation outcomes. The denaturation of the keratin in the outermost layer of skin is due to the irritation that happens due to the high amount of surfactant molecules. Hence, it is generally preferred to utilize surfactants that are identified based on low CMC concentration utilized and hence causes a low level of skin irritation.

4.6.3 BS as Prebiotics

The idea of prebiotics is well recognized in the functional food sector where different ingredients are utilized to excite the growth of beneficial probiotic microorganisms. Prebiotics are extensively contained of indigestible dietary fiber, which is derived from food processing of fruits, vegetables, and agro-industrial by-products [123]. Functional substances that are influencing beneficial microbial growth showed the therapeutic potential are defined as the "pharmabiotics" [124]. The prebiotic ingredient in the cosmetics industry is having a definite role in maintaining skin microflora.

The skin microbiota of the *Propionibacterium acne* can be regulated by the presence of beneficial bacteria such as *S. epidermidis* [125]. The biomedical potential of the BSs producing microorganisms adds to the benefits to the cosmetic and personal care product formulations (Fig. 4.1). A survey was carried out by the BS producing LAB, which is recognized as GRAS microorganisms [127]. Such microorganisms can be used as a potential prebiotic component in healthcare and cosmetic formulations. LAB-derived BSs displayed efficient antimicrobial and antibiofilm by crude surfactants, i.e. glycoproteins isolated from *Lactococcus lactis* [128].

BSs obtained from the *L. paracasei* A20 displayed inhibition of various human pathogens [111]. In certain cases, the antibiofilm activity is more prominent as compared to the antimicrobial activity [129]. In that report, it was found that glycoprotein rich BSs obtained from *Lactobacillus agilis* inhibited approximately 10–15% of the growth of human pathogens. Although, there was no inhibition of *E. coli* and *C. albicans* recorded in the study involved LAB derived BS. Sophorolipids also exhibited significant antimicrobials activity at a concentration of 5% (v/v) (Diaz [130]).

Kim et al. [131] also demonstrated the antimicrobial potential by sophorolipids derived from the *C. bombicola* to control *B. subtilis, S. xylosus, S. mutans,* and *P. acnes,* correspondingly. Furthermore, lipopeptides obtained from the *Propionibacterium freudenreichii* to control various fungal and bacterial pathogens [132]. Luna et al. [24] reported that BS of *Candida sphaerica* displayed antimicrobial and antiadhesive properties against various pathogens.

The BS obtained from the *Candida lipolytica* displayed inhibition of the various oral health-associated pathogens at a concentration of 12 mg/L [26, 27]. Also, poor inhibition was reported with *C. albicans, E. coli, P. aeruginosa, S. aureus,* and *S. epidermidis.* BSs can be incorporated in the cosmetic formulation under the scope

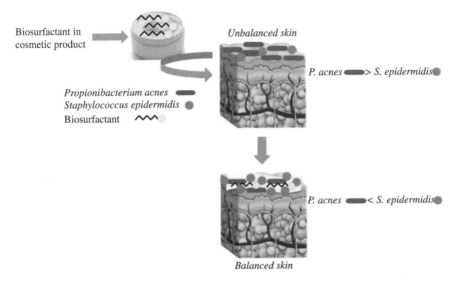

Fig. 4.1 Illustration of the prebiotic concept using biosurfactants in cosmetic products (Courtesy by: [126])

of the antimicrobial and emulsification agents as secondary claims as to the cosmetics regulation. Though, such antimicrobial behavior cannot be marketed in cosmetic formulations.

Glycolipids are among the best-studied BS utilized in cosmetics and healthcare preparations. For example, rhamnolipid based cosmetic preparations have been patented and utilized as anti-wrinkle and anti-aging materials. Rhamnolipids at a concentration of 0.001 up to 5% was found as an anti-aging cosmetics ointment [133]. MEL BSs showed significant antioxidant potential and superoxide scavenging assay with fibroblast cells [134]. Based on such an outcome, the report recommended that MELs have effective anti-aging constituents of skincare products. The report about the anti-aging potential of various BSs has been documented. Oligomeric BSs have been utilized as anti-aging creams, gel, and hair conditioning [135]. In a report, Kitagawa et al. [136] comprised a BS derived from *P. antarctica* in the preparation of a skincare product to enhance the skin quality such as skin roughness improvement. Rhamnolipids have been used as the key constituents in the formulation of hair shampoo (Desanto [137]). The rhamnolipids treatment displayed that the removal of the scalp, results in decrease of bad odor, and maintaining the hair shine for 3 days after treatment.

In the same line, it was demonstrated that hair shampoo containing sophorolipids in a synergistic combination with an anionic surfactant [138]. A personal healthcare product formulated containing glycolipid for skin washing [139]. BSs of *Starmella bombicola* and cocamidopropyl betaine in a ratio of 2:3 displayed appropriate properties [140]. Potentials of utilizing BS as a thickening material as related to the carbomer have been checked [141] where sulfosuccinates and BSs displayed significant thickening properties.

Antidandruff preparation along with BS and one oleic acid presents effective cosmetic solutions in shampoo, skin moisturizers, and body cleaning and gels [142]. A product containing MEL BS has been patented for the formulation of cosmetic pigments [136]. Product targeting oral health also has the potential to accommodate the BSs as a substitute to commonly used SDS [143]. Various researchers have emphasized the enormous potential of BSs for cosmetic preparations. Though, the quantity of information on their actual utilization of BS in creams, gels, lotions formulations, is very inadequate.

4.6.4   Healthcare Applications

An extensive range of bioactive microbial metabolites, together with BSs, are reported as having potential for healthcare and personal care formulations including biomedical applications. BSs obtained from *B. subtilis* were reported for the healing of wound infections in rats which results in epidermal regeneration [37–39]. The lipopeptides obtained from Bacillus genus with excellent free-radical scavenging potential have been reported. It help to control the inflammation and development of tissue formation, regeneration of epidermis cells [144]. At the same time, the BSs have been reported to control multidrug resistance bacteria [145] and display potential to control phytopathogenic fungal pathogens [146].

The glycolipid obtained from the *B. licheniformis* utilized for the preparation of ointments for the regeneration of the epithelial fibroblast cells along with rapid collagen deposition [147]. Lipopeptides derived from the *Acinetobacter junii* were reported for wound healing, antioxidant properties, and reduction of histopathological behavior [148]. Sophorolipids the other class of glycolipids reported from the yeast displayed significant antimicrobial potential to formulate antimicrobials to decrease the chances of the infection during wound healing [149]. Consequently, cytotoxic studies should be obligatory to explore any probable alteration on cells by their addition in healthcare or personal care formulations.

4.6.5   Oral Health

The oral microflora of the human and animals harbors an extensive and diverse range of microorganisms [150–152]. Various mechanisms have been identified as being responsible for this contribution to the normal status of health in the individual. Microorganisms associated with oral health can survive in a planktonic or biofilm development. Out of each planktonic cell easy to control as they are loosely anchored with oral surfaces and tissues, whereas the formation of dental plaque or biofilms protects the inhabitant which makes them less vulnerable to oral hygienic habits and bioactive agents [153]. A protein-rich fraction mainly present in the physiological saliva forms a substratum that permits microbial adhesion using specific receptors [154], which is subsequently trailed by co-aggregation and adhesion of various types of microbial colonizers for example Streptococci growth on the dental surface using EPS substances as an anchored for the biofilm development [155], which is extensively contained of sugars, proteins, lipids, and lipids [156]. Biofilm formation by Fusobacteria bridging colonization and co-aggregation of biofilm-forming cells or cell surface proteins provided that multi-binding sites for such microorganisms [157, 158]. The distinctive protection is also moderately related to the incidence of BSs producing microorganisms like *Streptococcus mitis*, which significantly control the adherence of a cariogenic bacterium, i.e. *Streptococcus mutans* [159].

Some of the BSs producer LAB involved in the maintenance of innate oral health Lactobacillus species [160, 161]. In general, BSs are found to be associated with innate oral health such as BSs derived from the *Streptococcus mitis* retard the adhesion of the *S. mutans*. In another study, the rhamnolipids obtained from the *Burkholderia thailandensis* demonstrated that it decreases the viability of various dental pathogens [162, 163].

The potential of the *Bacillus subtilis* lipopeptides incorporated in the dental toothpaste demonstrated effective antimicrobial potential against the Enterobacter sp. and *Salmonella typhimurium* and other multidrug-resistant pathogens [145, 164].

The BSs examined are capable as a probable substitute for synthetic surfactants in oral health care products and antimicrobials. The synergy of BSs, chitosan, and sodium fluoride demonstrating the positive amalgamation between the substances. The BSs based dental formulations are non-toxic, which recommends their effective and safe use. The in vitro demonstration of such data provides an elementary and fundamental way ahead to give endurance to such studies by showing such studies in biological systems.

4.6.6   Skincare Formulations

The majority of the antimicrobial preservatives used in skincare and personal care products demonstrated skin allergies and skin irritation. The antibacterial preservatives used in the majority of personal care products are chemical-based surfactants by dealings with keratin, elastin, or collagen and also inspire the exclusion of lipids from the skin cells [165]. But at the same time, BSs based preparations are contained in lipid and proteins and are biocompatible with the skin epidermal cells and tissues [166]. Probiotic lactobacilli microorganisms have the additional benefits of being non-toxic, eco-degradable, and eco-friendly [161]. BSs obtained from the *Lactobacillus pentosus* act against various skin microflora Vecino et al. [126]. Sharma and Saharan [1] demonstrated the antimicrobial potential of BSs obtained from *Lactobacillus helveticus* and exhibited significant potential against *E. coli* and *S. epidermidis*. While Gudiña et al. [129] demonstrating the BSs of *L. agilis* exhibits antimicrobial potential against *E. coli* and *C. albicans*.

Sopholipids obtained from the yeast showed improved antimicrobial activity against the acne-causing *Propionibacterium acnes* [167]. In one case, rhamnolipids enable the production of the proteins via skin protein flagellin [168]. Flagellin activates the keratinocytes cells to initiate the production of the antimicrobial protein which can control and can kill *P. aeruginosa*. BSs containing hydrogels showed a defense mechanism to control multidrug resistance skin pathogens [169]. Such behavior of the BSs suggests the efficacy against skin infections and pathogens.

4.6.7   Drawbacks and Future Trends

Various efforts have been made in previous years to utilize the BSs in the food and healthcare sector. Though, for their effective exploration, it is authoritative to emphasize in vivo toxicity assessment. The major drawback for the adoption of BS is the concern about their safety. Utilization of the BSs obtained from the non-pathogenic strains, such as LAB, yeasts, and various probiotics strains is a significant alteration to reduce toxicity can be an excellent approach to tackle such a problem. The greener BSs may present an advantage against synthetic emulsifiers, particularly for explicit markets like organic, and non-GMO, and Kosher certified food. Microbial BSs are not homogeneously pure but on the other hand, chemical-based surfactants are pure homogenous substances such as SDS and food emulsifiers. The impurity of BSs leads to the problem of non-specific properties. At the same time, the emergence of the resistance to the chemicals-based preservatives and surfactants can be minimized using microbial surfactants. Microbial surfactants do not lead to any resistance as they are non-specific in killing mechanisms. Another drawback or challenges associated with the utilization of the food is the sensorial effect of the incorporation of the BSs in food preparations and their interaction with food constituents should be examined. Because of the vast microbial diversity, there is a huge demand to explore novel surfactants, particularly from extremophiles. The search for the new metabolites coupled with metabolic engineering approaches may come out in competitive surfactants. An advanced trend is the utilization of BSs as mockups for the tailor-made of green surfactants, like an enzymatic synthesis of the glycolipids.

Moreover, the regulation of the governing bodies about the food ingredients needs considerable research and risk assessment to include approvals for food inclusions. Though, the commercialization is limited due to its difficulty in product recovery, and low yields, all of which impact the production cost when compared to the synthetic surfactants. BSs with antimicrobial activity with low MIC value must be explored only for the food preservatives. To rule out the safety concerns and toxicity concerns, we should select novel strains with GRAS status or non-pathogenic BSs producers.

The exclusive properties existing by microbial origin BSs recommend that they can, in the coming future, BSs can be a part of the food formulations and processing chain, even as an active food additive, and cleaning solutions.

References

1. Sharma D, Saharan BS (2016) Functional characterization of biomedical potential of biosurfactant produced by Lactobacillus helveticus. Biotechnol Rep 11:27–35
2. De Almeida DG, Soares Da Silva RDCF, Luna JM, Rufino RD, Santos VA, Banat IM, Sarubbo LA (2016) Biosurfactants: promising molecules for petroleum biotechnology advances. Front Microbiol 7:1718
3. Kumar P, Sharma PK, Sharma PK, Sharma D (2015) Micro-algal lipids: a potential source of biodiesel. J Innov Pharm Biol Sci 2(2):135–143
4. Campos JM, Stamford TLM, Sarubbo LA (2019) Characterization and application of a biosurfactant isolated from Candida utilis in salad dressings. Biodegradation 30(4):313–324
5. Sharma D, Gupta E, Singh J, Vyas P, Dhanjal DS (2018) Microbial biosurfactants in food sanitation. In: Sustainable food systems from agriculture to industry. Academic Press, pp 341–368
6. Csáki KF (2011) Synthetic surfactant food additives can cause intestinal barrier dysfunction. Med Hypotheses 76(5):676–681
7. Paterson BM, Turner JR (2008) Inhibitors of intestinal barrier dysfunction. In: Frontiers in celiac disease, vol 12. Karger Publishers, Basel, pp 157–171
8. Sharma D, Saharan BS, Kapil S (2016) Biosurfactants of lactic acid bacteria. Springer, Cham
9. Nitschke M, Costa SGVAO (2014) Biosurfactants in the Food Industry. In: Biosurfactants research trends and applications. CRC Press, Boca Raton, pp 177–196
10. Sharma D, Saharan BS (eds) (2018) Microbial cell factories. CRC Press
11. Bridier A, Sanchez-Vizuete P, Guilbaud M, Piard JC, Naitali M, Briandet R (2015) Biofilm-associated persistence of food-borne pathogens. Food Microbiol 45:167–178
12. Srey S, Jahid IK, Ha SD (2013) Biofilm formation in food industries: a food safety concern. Food Control 31(2):572–585
13. Cappitelli F, Polo A, Villa F (2014) Biofilm formation in food processing environments is still poorly understood and controlled. Food Eng Rev 6(1–2):29–42
14. Macia MD, Rojo-Molinero E, Oliver A (2014) Antimicrobial susceptibility testing in biofilm-growing bacteria. Clin Microbiol Infect 20(10):981–990
15. Olszewska MA (2013) Microscopic findings for the study of biofilms in food environments. Acta Biochim Pol 60(4)
16. Araujo LVD, Freire DMG, Nitschke M (2013) Biosurfactants: anticorrosive, antibiofilm and antimicrobial properties. Quim Nova 36(6):848–858
17. Branen AL, Haggerty RJ (2002) Introduction to food additives. In: Food additives. Kanopy Streaming, San Francisco, pp 1–9
18. Pandey RM, Upadhyay SK (2012) Food additive division of genetics. In: Plant breeding & Agrotechnology, National Botanical Research Institute, Lucknow, India. Doi, 10, 34455
19. Transparency Market Research (2016) Global food additives market to reach US$39.85 by 2021, propelled by increasing demand for healthier food additives. Available from http://www.transparency marketresearch.com/pressrelease/food-additives.html. Accessed June 24 2016
20. Kuiper I, Lagendijk EL, Pickford R, Derrick JP, Lamers GE, Thomas-Oates JE et al (2004) Characterization of two Pseudomonas putida lipopeptide biosurfactants, putisolvin I and II, which inhibit biofilm formation and break down existing biofilms. Mol Microbiol 51 (1):97–113
21. Janek T, Łukaszewicz M, Rezanka T, Krasowska A (2010) Isolation and characterization of two new lipopeptide biosurfactants produced by Pseudomonas fluorescens BD5 isolated from water from the Arctic archipelago of Svalbard. Bioresour Technol 101(15):6118–6123
22. Rivardo F, Turner RJ, Allegrone G, Ceri H, Martinotti MG (2009) Anti-adhesion activity of two biosurfactants produced by Bacillus spp. prevents biofilm formation of human bacterial pathogens. Appl Microbiol Biotechnol 83(3):541–553

23. Pradhan AK, Pradhan N, Mall G, Panda HT, Sukla LB, Panda PK, Mishra BK (2013) Application of lipopeptide biosurfactant isolated from a halophile: Bacillus tequilensis CH for inhibition of biofilm. Appl Biochem Biotechnol 171(6):1362–1375

24. Luna JM, Rufino RD, Sarubbo LA, Rodrigues LR, Teixeira JA, de Campos-Takaki GM (2011) Evaluation antimicrobial and antiadhesive properties of the biosurfactant Lunasan produced by Candida sphaerica UCP 0995. Curr Microbiol 62(5):1527–1534

25. Dusane DH, Dam S, Nancharaiah YV, Kumar AR, Venugopalan VP, Zinjarde SS (2012) Disruption of Yarrowia lipolytica biofilms by rhamnolipid biosurfactant. Aquat Biosyst 8 (1):1–7

26. Rufino RD, de Luna JM, Sarubbo LA, Rodrigues LRM, Teixeira JAC, de Campos-Takaki GM (2011a) Antimicrobial and anti-adhesive potential of a biosurfactants produced by Candida species. In: Practical applications in biomedical engineering. InTech, Rijeka

27. Rufino RD, Luna JM, Sarubbo LA, Rodrigues LRM, Teixeira JAC, Campos-Takaki GM (2011b) Antimicrobial and anti-adhesive potential of a biosurfactant Rufisan produced by Candida lipolytica UCP 0988. Colloids Surf B: Biointerfaces 84(1):1–5

28. Dusane DH, Pawar VS, Nancharaiah YV, Venugopalan VP, Kumar AR, Zinjarde SS (2011) Anti-biofilm potential of a glycolipid surfactant produced by a tropical marine strain of Serratia marcescens. Biofouling 27(6):645–654

29. Benincasa M, Abalos A, Oliveira I, Manresa A (2004) Chemical structure, surface properties and biological activities of the biosurfactant produced by Pseudomonas aeruginosa LBI from soapstock. Antonie Van Leeuwenhoek 85(1):1–8

30. Kim BS, Lee JY, Hwang BK (2000) In vivo control and in vitro antifungal activity of rhamnolipid B, a glycolipid antibiotic, against Phytophthora capsici and Colletotrichum orbiculare. Pest Manag Sci Formerly Pest Sci 56(12):1029–1035

31. Haba E, Pinazo A, Jauregui O, Espuny MJ, Infante MR, Manresa A (2003) Physicochemical characterization and antimicrobial properties of rhamnolipids produced by Pseudomonas aeruginosa 47T2 NCBIM 40044. Biotechnol Bioeng 81(3):316–322

32. Varnier AL, Sanchez L, Vatsa P, Boudesocque L, Garcia-Brugger A, Rabenoelina F et al (2009) Bacterial rhamnolipids are novel MAMPs conferring resistance to Botrytis cinerea in grapevine. Plant Cell Environ 32(2):178–193

33. Nitschke M, Costa SG, Contiero J (2010) Structure and applications of a rhamnolipid surfactant produced in soybean oil waste. Appl Biochem Biotechnol 160(7):2066–2074

34. Sotirova A, Spasova D, Vasileva-Tonkova E, Galabova D (2009) Effects of rhamnolipid-biosurfactant on cell surface of Pseudomonas aeruginosa. Microbiol Res 164(3):297–303

35. Chander CS, Lohitnath T, Kumar DM, Kalaichelvan PT (2012) Production and characterization of biosurfactant from Bacillus subtilis MTCC441 and its evaluation to use as bioemulsifier for food bio-preservative. Adv Appl Sci Res 3(3):1827–1831

36. Campos JM, Stamford TL, Rufino RD, Luna JM, Stamford TCM, Sarubbo LA (2015) Formulation of mayonnaise with the addition of a bioemulsifier isolated from Candida utilis. Toxicol Rep 2:1164–1170

37. Zouari R, Besbes S, Ellouze-Chaabouni S, Ghribi-Aydi D (2016b) Cookies from composite wheat–sesame peels flours: dough quality and effect of Bacillus subtilis SPB1 biosurfactant addition. Food Chem 194:758–769

38. Zouari R, Hamden K, El Feki A, Chaabouni K, Makni-Ayadi F, Kallel C et al (2016c) Protective and curative effects of Bacillus subtilis SPB1 biosurfactant on high-fat-high-fructose diet induced hyperlipidemia, hypertriglyceridemia and deterioration of liver function in rats. Biomed Pharmacother 84:323–329

39. Zouari R, Moalla-Rekik D, Sahnoun Z, Rebai T, Ellouze-Chaabouni S, Ghribi-Aydi D (2016a) Evaluation of dermal wound healing and in vitro antioxidant efficiency of Bacillus subtilis SPB1 biosurfactant. Biomed Pharmacother 84:878–891

40. Haesendonck IV, Vanzeveren E (2006) Rhamnolipids in bakery products. US 20060233935 A1

41. Trummler K, Effenberger F, Syldatk C (2003) An integrated microbial/enzymatic process for production of rhamnolipids and L-(+)-rhamnose from rapeseed oil with Pseudomonas sp. DSM 2874. Eur J Lipid Sci Technol 105(10):563–571

42. Kiran GS, Priyadharsini S, Sajayan A, Priyadharsini GB, Poulose N, Selvin J (2017) Production of lipopeptide biosurfactant by a marine Nesterenkonia sp and its application in food industry. Frontiers Microbiol 8:1138

43. Mnif I, Besbes S, Ellouze R, Ellouze-Chaabouni S, Ghribi D (2012) Improvement of bread quality and bread shelf-life by Bacillus subtilis biosurfactant addition. Food Sci Biotechnol 21 (4):1105–1112

44. Silva IA, Veras BO, Ribeiro BG, Aguiar JS, Guerra JMC, Luna JM, Sarubbo LA (2020) Production of cupcake-like dessert containing microbial biosurfactant as an emulsifier. PeerJ 8:e9064

45. Vasudevan S, Prabhune AA (2018) Photophysical studies on curcumin-sophorolipid nanostructures: applications in quorum quenching and imaging. R Soc Open Sci 5(2):170865

46. Konkol D, Szmigiel I, Domżał-Kędzia M, Kułażyński M, Krasowska A, Opaliński S et al (2019) Biotransformation of rapeseed meal leading to production of polymers, biosurfactants, and fodder. Bioorg Chem 93:102865

47. Song B, Zhu W, Song R, Yan F, Wang Y (2019) Exopolysaccharide from Bacillus vallismortis WF4 as an emulsifier for antifungal and antipruritic peppermint oil emulsion. Int J Biol Macromol 125:436–444

48. Campos JM, Stamford TL, Sarubbo LA (2014) Production of a bioemulsifier with potential application in the food industry. Appl Biochem Biotechnol 172(6):3234–3252

49. Nitschke M, Costa SG, Contiero J (2005) Rhamnolipid surfactants: an update on the general aspects of these remarkable biomolecules. Biotechnol Prog 21(6):1593–1600

50. Nitschke M, Costa SGVAO (2007) Biosurfactants in food industry. Trends Food Sci Technol 18(5):252–259

51. Lima ÁS, Alegre RM (2009) Evaluation of emulsifier stability of biosurfactant produced by Saccharomyces lipolytica CCT-0913. Braz Arch Biol Technol 52(2):285–290

52. Bai L, McClements DJ (2016) Formation and stabilization of nanoemulsions using biosurfactants: Rhamnolipids. J Colloid Interface Sci 479:71–79

53. Ely CM (1951) Chick-growth stimulation produced by surfactants. Science (Washington) 114:523–524

54. Kralova I, Sjöblom J (2009) Surfactants used in food industry: a review. J Dispers Sci Technol 30(9):1363–1383

55. Stauffer CE, Ambigaipalan P, Shahidi F (2005) Emulsifiers for the food industry. In: Bailey's industrial oil and fat products. Wiley-Interscience, Hoboken, pp 1–36

56. McClements DJ, Gumus CE (2016) Natural emulsifiers—biosurfactants, phospholipids, biopolymers, and colloidal particles: molecular and physicochemical basis of functional performance. Adv Colloid Interf Sci 234:3–26

57. Gandhi NR, Skebba V, Takemoto J, Bensaci M. "Antimycotic rhamnolipid compositions and related methods of use." U.S. Patent Application 11/351,572, filed August 16, 2007

58. Van Haesendonck IPH, Vanzeveren ECA (2004) Rhamnolipids in bakery products. *International Application Patent (PCT) WO*, 2004-040984

59. Suresh Chander CR, Lohitnath T, Mukesh Kumar DJ, Kalaichelvan PT (2012) Production and characterization of biosurfactant from bacillus subtilis MTCC441 and its evaluation to use as bioemulsifier for food bio–preservative. Adv Appl Sci Res 3(3):1827–1831

60. Uzoigwe C, Burgess JG, Ennis CJ, Rahman PK (2015) Bioemulsifiers are not biosurfactants and require different screening approaches. Front Microbiol 6:245

61. Sagalowicz L, Leser ME (2010) Delivery systems for liquid food products. Curr Opin Colloid Interface Sci 15(1–2):61–72

62. Manoharan MJ, Bradeeba K, Parthasarathi R, Sivakumaar PK, Chauhan PS, Tipayno S, Benson A, Sa T (2012) Development of surfactin-based nanoemulsion formulation from

selected cooking oils: evaluation for antimicrobial activity against selected food associated microorganisms. J Taiwan Inst Chem E 43:172–180

63. Farheen V, Saha SB, Pyne S, Chowdhury BR (2016) Production of nanobiosurfactant from Pseudomonas aeruginosa and its application in bakery industry. Int J Adv Res Biol Eng Sci Technol 2:67(SI8)

64. Khoshdast H, Sam A (2012) An efficiency evaluation of iron concentrates flotation using rhamnolipid biosurfactant as a frothing reagent. Environ Eng Res 17(1):9–15

65. Vaughn SF, Behle RW, Skory CD, Kurtzman CP, Price NPJ (2014) Utilization of sophorolipids as biosurfactants for postemergence herbicides. Crop Prot 59:29–34

66. Gudiña EJ, Rangarajan V, Sen R, Rodrigues LR (2013) Potential therapeutic applications of biosurfactants. Trends Pharmacol Sci 34(12):667–675

67. Randu M, Sylvie HERY, Ravier, P, Deprey S (2019) U.S. patent application no. 16/085,348

68. Sekhar KP, Nayak RR (2018) Nonionic glycolipids for chromium flotation-and emulsion (W/O and O/W)-based bioactive release. Langmuir 34(47):14347–14357

69. De Zoysa GH, Glossop HD, Sarojini V (2018) Unexplored antifungal activity of linear battacin lipopeptides against planktonic and mature biofilms of C. albicans. Eur J Med Chem 146:344–353

70. Braun AC, Ilko D, Merget B, Gieseler H, Germershaus O, Holzgrabe U, Meinel L (2015) Predicting critical micelle concentration and micelle molecular weight of polysorbate 80 using compendial methods. Eur J Pharm Biopharm 94:559–568

71. Goodarzi F, Zendehboudi S (2019) Effects of salt and surfactant on interfacial characteristics of water/oil systems: molecular dynamic simulations and dissipative particle dynamics. Ind Eng Chem Res 58(20):8817–8834

72. Singh AK, Rautela R, Cameotra SS (2014) Substrate dependent in vitro antifungal activity of Bacillus sp strain AR2. Microb Cell Factories 13(1):67

73. Israelachvili JN (2011) Intermolecular and surface forces. Academic press, London

74. Kitamoto D, Morita T, Fukuoka T, Konishi MA, Imura T (2009) Self-assembling properties of glycolipid biosurfactants and their potential applications. Curr Opin Colloid Interface Sci 14 (5):315–328

75. Worakitkanchanakul W, Imura T, Fukuoka T, Morita T, Sakai H, Abe M et al (2008) Aqueous-phase behavior and vesicle formation of natural glycolipid biosurfactant, mannosylerythritol lipid-B. Colloids Surf B: Biointerfaces 65(1):106–112

76. Nguyen TT, Edelen A, Neighbors B, Sabatini DA (2010) Biocompatible lecithin-based microemulsions with rhamnolipid and sophorolipid biosurfactants: formulation and potential applications. J Colloid Interface Sci 348(2):498–504

77. Liu F, Zhu Z, Ma C, Luo X, Bai L, Decker EA et al (2016) Fabrication of concentrated fish oil emulsions using dual-channel microfluidization: impact of droplet concentration on physical properties and lipid oxidation. J Agric Food Chem 64(50):9532–9541

78. Kiran GS, Selvin J, Manilal A, Sujith S (2011) Biosurfactants as green stabilizers for the biological synthesis of nanoparticles. Crit Rev Biotechnol 31(4):354–364

79. Palanisamy P, Raichur AM (2009) Synthesis of spherical NiO nanoparticles through a novel biosurfactant mediated emulsion technique. Mater Sci Eng C 29(1):199–204

80. Xie Y, Ye R, Liu H (2006) Synthesis of silver nanoparticles in reverse micelles stabilized by natural biosurfactant. Colloids Surf A Physicochem Eng Asp 279(1–3):175–178

81. Reddy AS, Chen CY, Baker SC, Chen CC, Jean JS, Fan CW et al (2009) Synthesis of silver nanoparticles using surfactin: A biosurfactant as stabilizing agent. Mater Lett 63 (15):1227–1230

82. Biswas M, Raichur AM (2008) Electrokinetic and rheological properties of nano zirconia in the presence of rhamnolipid biosurfactant. J Am Ceram Soc 91(10):3197–3201

83. Shim GY, Kim SH, Han SE, Kim YB, Oh YK (2009) Cationic surfactin liposomes for enhanced cellular delivery of siRNA. Asian J Pharmaceut Sci 4:207–214

84. Igarashi S, Hattori Y, Maitani Y (2006) Biosurfactant MEL-A enhances cellular association and gene transfection by cationic liposome. J Control Release 112(3):362–368

85. Imura T, Yanagishita H, Ohira J, Sakai H, Abe M, Kitamoto D (2005) Thermodynamically stable vesicle formation from glycolipid biosurfactant sponge phase. Colloids Surf B: Biointerfaces 43(2):115–121

86. Inoh Y, Kitamoto D, Hirashima N, Nakanishi M (2001) Biosurfactants of MEL-A increase gene transfection mediated by cationic liposomes. Biochem Biophys Res Commun 289 (1):57–61

87. Cheow WS, Hadinoto K (2012) Lipid-polymer hybrid nanoparticles with rhamnolipid-triggered release capabilities as anti-biofilm drug delivery vehicles. Particuology 10 (3):327–333

88. Gaur VK, Regar RK, Dhiman N, Gautam K, Srivastava JK, Patnaik S et al (2019) Biosynthesis and characterization of sophorolipid biosurfactant by Candida spp.: Application as food emulsifier and antibacterial agent. Bioresour Technol 285:121314

89. D'Auria L, Deleu M, Dufour S, Mingeot-Leclercq MP, Tyteca D (2013) Surfactins modulate the lateral organization of fluorescent membrane polar lipids: a new tool to study drug: membrane interaction and assessment of the role of cholesterol and drug acyl chain length. Biochimica et Biophysica Acta (BBA)-Biomembranes 1828(9):2064–2073

90. Huang X, Suo J, Cui Y (2011) Optimization of antimicrobial activity of surfactin and polylysine against Salmonella enteritidis in milk evaluated by a response surface methodology. Foodborne Pathog Dis 8(3):439–443

91. Meng Q, Zhang G (2012) Application of rhamnolipid as biological cleaning agent. Chinese Patent CN 102399644

92. Sambanthamoorthy K, Feng X, Patel R, Patel S, Paranavitana C (2014) Antimicrobial and antibiofilm potential of biosurfactants isolated from lactobacilli against multi-drug-resistant pathogens. BMC Microbiol 14(1):197

93. Hirata Y, Ryu M, Oda Y, Igarashi K, Nagatsuka A, Furuta T, Sugiura M (2009) Novel characteristics of sophorolipids, yeast glycolipid biosurfactants, as biodegradable low-foaming surfactants. J Biosci Bioeng 108(2):142–146

94. Pierce D, Heilman TJ (1998) Germicidal composition. World Patent 9816192

95. Jing C, Bingbing Y (2010) Sophorolipid fruit preservative and use thereof in fruit preservation. Chinese Patent CN, 101886047

96. Ortiz A, Teruel JA, Espuny MJ, Marqués A, Manresa Á, Aranda FJ (2008) Interactions of a Rhodococcus sp. biosurfactant trehalose lipid with phosphatidylethanolamine membranes. Biochimica et Biophysica Acta (BBA)-Biomembranes 1778(12):2806–2813

97. Singh P, Patil Y, Rale V (2019) Biosurfactant production: emerging trends and promising strategies. J Appl Microbiol 126(1):2–13

98. Zaragoza A, Aranda FJ, Espuny MJ, Teruel JA, Marques A, Manresa A, Ortiz A (2009) Mechanism of membrane permeabilization by a bacterial trehalose lipid biosurfactant produced by Rhodococcus sp. Langmuir 26:8567–8572

99. De Araujo LV, Abreu F, Lins U, Santa Anna LMDM, Nitschke M, Freire DMG (2011) Rhamnolipid and surfactin inhibit Listeria monocytogenes adhesion. Food Res Int 44(1):481–488

100. Magalhães L, Nitschke M (2013) Antimicrobial activity of rhamnolipids against Listeria monocytogenes and their synergistic interaction with nisin. Food Control 29(1):138–142

101. Maukonen J, Mättö J, Wirtanen G, Raaska L, Mattila-Sandholm T, Saarela M (2003) Methodologies for the characterization of microbes in industrial environments: a review. J Ind Microbiol Biotechnol 30(6):327–356

102. Sharma D, Saharan BS, Chauhan N, Bansal A, Procha S (2014) Production and structural characterization of Lactobacillus helveticus derived biosurfactant. Sci World J:2014

103. Bagge-Ravn D, Ng Y, Hjelm M, Christiansen JN, Johansen C, Gram L (2003) The microbial ecology of processing equipment in different fish industries—analysis of the microflora during processing and following cleaning and disinfection. Int J Food Microbiol 87(3):239–250

104. Larsen N, Thorsen L, Kpikpi EN, Stuer-Lauridsen B, Cantor MD, Nielsen B et al (2014) Characterization of Bacillus spp. strains for use as probiotic additives in pig feed. Appl Microbiol Biotechnol 98(3):1105–1118

105. Sharma BK, Saha A, Rahaman L, Bhattacharjee S, Tribedi P (2015) Silver inhibits the biofilm formation of Pseudomonas aeruginosa. Adv Microbiol 5(10):677

106. Rautela R, Cameotra SS (2014) Role of biopolymers in industries: their prospective future applications. In: Environment and Sustainable Development. Springer, New Delhi, pp 133–142

107. Meylheuc T, Van Oss CJ, Bellon-Fontaine MN (2001) Adsorption of biosurfactant on solid surfaces and consequences regarding the bioadhesion of Listeria monocytogenes LO28. J Appl Microbiol 91(5):822–832

108. Kim H, Ryu JH, Beuchat LR (2006) Attachment of and biofilm formation by Enterobacter sakazakii on stainless steel and enteral feeding tubes. Appl Environ Microbiol 72 (9):5846–5856

109. Meylheuc T, Renault M, Bellon-Fontaine MN (2006) Adsorption of a biosurfactant on surfaces to enhance the disinfection of surfaces contaminated with Listeria monocytogenes. Int J Food Microbiol 109(1–2):71–78

110. Nitschke M, Araújo LV, Costa SGVAO, Pires RC, Zeraik AE, Fernandes ACLB et al (2009) Surfactin reduces the adhesion of food-borne pathogenic bacteria to solid surfaces. Lett Appl Microbiol 49(2):241–247

111. Gudina EJ, Teixeira JA, Rodrigues LR (2010) Isolation and functional characterization of a biosurfactant produced by Lactobacillus paracasei. Colloids Surf B: Biointerfaces 76 (1):298–304

112. McLandsborough L, Rodriguez A, Pérez-Conesa D, Weiss J (2006) Biofilms: at the interface between biophysics and microbiology. Food Biophysics 1(2):94–114

113. Verbeke W (2006) Functional foods: consumer willingness to compromise on taste for health? Food Qual Prefer 17(1–2):126–131

114. Ozdener MH, Ashby RD, Jyotaki M, Elkaddi N, Spielman AI, Bachmanov AA, Solaiman DK (2019) Sophorolipid biosurfactants activate taste receptor type 1 member 3-mediated taste responses and block responses to bitter taste in vitro and in vivo. J Surfactant Deterg 22 (3):441–449

115. Kaboré D, Gagnon M, Roy D, Sawadogo-Lingani H, Diawara B, LaPointe G (2018) Rapid screening of starter cultures for maari based on antifungal properties. Microbiol Res 207:66–74

116. Mcclements DJ (2007) Critical review of techniques and methodologies for characterization of emulsion stability. Crit Rev Food Sci Nutr 47(7):611–649

117. Zhang K, Su L, Wu J (2020) Recent advances in recombinant protein production by Bacillus subtilis. Annu Rev Food Sci Technol 11:295–318

118. Partal P, Guerrero A, Berjano M, Gallegos C (1999) Transient flow of o/w sucrose palmitate emulsions. J Food Eng 41(1):33–41

119. Saravanakumari P, Mani K (2010) Structural characterization of a novel xylolipid biosurfactant from Lactococcus lactis and analysis of antibacterial activity against multi-drug resistant pathogens. Bioresour Technol 101(22):8851–8854

120. Marchant R, Banat IM (2012) Biosurfactants: a sustainable replacement for chemical surfactants? Biotechnol Lett 34(9):1597–1605

121. Food and Drug Administration (2010) Substances generally recognized as safe; reopening of the comment period. Fed Regist 75:81536

122. Konishi M, Imura T, Morita T, Fukuoka T, Kitamoto D (2007) A yeast glycolipid biosurfactant, mannosyl-erythritol lipid, shows high binding affinity towards lectins on a self-assembled monolayer system. Biotechnol Lett 29:473–480

123. Dominguez AL, Rodrigues LR, Lima NM, Teixeira JA (2014) An overview of the recent developments on fructooligosaccharide production and applications. Food Bioprocess Technol 7(2):324–337

124. Hutkins RW, Krumbeck JA, Bindels LB, Cani PD, Fahey G Jr, Goh YJ, Hamaker B et al (2016) Prebiotics: why definitions matter. Curr Opin Biotechnol 37:1–7

125. Simmering R, Breves R (2010) Prebiotic cosmetics. In: Nutrition for healthy skin. Springer, Berlin, pp 137–147

126. Vecino X, Rodríguez-López L, Ferreira D, Cruz JM, Moldes AB, Rodrigues LR (2018) Bioactivity of glycolipopeptide cell-bound biosurfactants against skin pathogens. Int J Biol Macromol 109:971–979

127. Bourdichon F, Berger B, Casaregola S, Farrokh C, Frisvad JC, Gerds ML et al (2012) Safety demonstration of microbial food cultures (MFC) in fermented food products, vol 455. FIL/IDF, Brussels, p 2

128. Rodrigues LR, Teixeira JA, van der Mei HC, Oliveira R (2006) Physicochemical and functional characterization of a biosurfactant produced by Lactococcus lactis 53. Colloids Surf B: Biointerfaces 49(1):79–86

129. Gudiña EJ, Fernandes EC, Teixeira JA, Rodrigues LR (2015) Antimicrobial and anti-adhesive activities of cell-bound biosurfactant from Lactobacillus agilis CCUG31450. RSC Adv 5 (110):90960–90968

130. De Rienzo MAD, Banat IM, Dolman B, Winterburn J, Martin PJ (2015) Sophorolipid biosurfactants: possible uses as antibacterial and antibiofilm agent. New Biotechnol 32 (6):720–726

131. Kim K, Dalsoo Y, Youngbum K, Baekseok L, Doonhoon S, Eun-Ki KIM (2002) Characteristics of sophorolipid as an antimicrobial agent. J Microbiol Biotechnol 12 (2):235–241

132. Hajfarajollah H, Mokhtarani B, Noghabi KA (2014) Newly antibacterial and antiadhesive lipopeptide biosurfactant secreted by a probiotic strain, Propionibacterium freudenreichii. Appl Biochem Biotechnol 174(8):2725–2740

133. Piljac T, Piljac G 1999 Use of rhamnolipids in wound healing, treating burn shock, athero-sclerosis, organ transplants, depression, schizophrenia and cosmetics European Patent EP1889623 A3, February 20

134. Takahashi M, Morita T, Fukuoka T, Imura T, Kitamoto D (2012) Glycolipid biosurfactants, mannosylerythritol lipids, show antioxidant and protective effects against H2O2-induced oxidative stress in cultured human skin fibroblasts. J Oleo Sci 61(8):457–464

135. Owen D, Fan L (2013) *U.S. Patent No. 8,586,541*. Washington, DC: U.S. Patent and Trademark Office

136. Kitagawa M, Suzuki M, Yamamoto S, Sogabe A, Kitamoto D, Imura T, et al. (2008). Skin care cosmetic and skin and agent for preventing skin roughness containing biosurfactants. Patent EP, 1964546, A1

137. Desanto K (2008) Rhamnolipid-based formulations. World Patent WO 2008013899 A3

138. Trevor F, Crawford R, Garry L, et al. (2013) Mild to the skin, foaming detergent composition. Patent US 8563490 B2

139. Parry NJ, Stevenson PS. Personal care compositions. Patent WO 2014095367 A1; 2014

140. Kulkarni S, Choudhary P (2011) Production and isolation of biosurfactant-sophorolipid and its application in body wash formulation. Asian J Microbiol Biotechnol Environ Sci 13:217–221

141. Schwab P, Kortemeier U, Hartung C et al (2014) Cosmetic formulation containing copolymer and sulfosuccinate and/or biosurfactant. Patent WO 2014166796 A1

142. Allef P, Hartung C, Schilling M (2016) *U.S. Patent No. 9,271,908*. Washington, DC: U.S. Patent and Trademark Office

143. Das I, Roy S, Chandni S, Karthik L, Kumar G, Rao KVB (2013) Biosurfactant from marine actinobacteria and its application in cosmetic formulation of toothpaste. Pharm Lett 5(5):1–6

144. Jemil N, Ayed HB, Manresa A, Nasri M, Hmidet N (2017) Antioxidant properties, antimicro-bial and anti-adhesive activities of DCS1 lipopeptides from Bacillus methylotrophicus DCS1. BMC Microbiol 17(1):144

145. Ghribi D, Abdelkefi-Mesrati L, Mnif I, Kammoun R, Ayadi I, Saadaoui I et al (2012) Investigation of antimicrobial activity and statistical optimization of Bacillus subtilis SPB1 biosurfactant production in solid-state fermentation. J Biomed Biotechnol 2012
146. Mnif I, Ghribi D (2015) Potential of bacterial derived biopesticides in pest management. Crop Prot 77:52–64
147. Gupta S, Raghuwanshi N, Varshney R, Banat IM, Srivastava AK, Pruthi PA, Pruthi V (2017) Accelerated in vivo wound healing evaluation of microbial glycolipid containing ointment as a transdermal substitute. Biomed Pharmacother 94:1186–1196
148. Ohadi M, Forootanfar H, Rahimi HR, Jafari E, Shakibaie M, Eslaminejad T, Dehghannoudeh G (2017) Antioxidant potential and wound healing activity of biosurfactant produced by Acinetobacter junii B6. Curr Pharm Biotechnol 18(11):900–908
149. Lydon HL, Baccile N, Callaghan B, Marchant R, Mitchell CA, Banat IM (2017) Adjuvant antibiotic activity of acidic sophorolipids with potential for facilitating wound healing. Antimicrob Agents Chemother 61(5)
150. Marsh PD (2000) Role of the oral microflora in health. Microb Ecol Health Dis 12(3):130–137
151. Marsh PD (2010) Microbiology of dental plaque biofilms and their role in oral health and caries. Dental Clinics 54(3):441–454
152. Wright JT, Graham F, Hayes C, Ismail AI, Noraian KW, Weyant RJ et al (2013) A systematic review of oral health outcomes produced by dental teams incorporating midlevel providers. J Am Dent Assoc 144(1):75–91
153. Socransky SS, Haffajee AD (2002) Dental biofilms: difficult therapeutic targets. Periodontology 2000(28):12–55
154. Kreth J, Merritt J, Qi F (2009) Bacterial and host interactions of oral streptococci. DNA Cell Biol 28(8):397–403
155. Kolenbrander PE, Palmer RJ, Periasamy S, Jakubovics NS (2010) Oral multispecies biofilm development and the key role of cell–cell distance. Nat Rev Microbiol 8(7):471–480
156. Flemming HC, Wingender J (2010) The biofilm matrix. Nat Rev Microbiol 8(9):623–633
157. Tao R, Jurevic RJ, Coaulton KK, Tsutsui MT, Roberts MC, Kimball JR et al (2005) Salivary antimicrobial peptide expression and dental caries experience in children. Antimicrob Agents Chemother 49(9):3883–3888
158. He X, Hu W, Kaplan CW, Guo L, Shi W, Lux R (2012) Adherence to streptococci facilitates Fusobacterium nucleatum integration into an oral microbial community. Microb Ecol 63 (3):532–542
159. Reid G, Younes JA, Van der Mei HC, Gloor GB, Knight R, Busscher HJ (2011) Microbiota restoration: natural and supplemented recovery of human microbial communities. Nat Rev Microbiol 9(1):27–38
160. Haukioja A (2010) Probiotics and oral health. European journal of dentistry 4(3):348
161. Satpute SK, Kulkarni GR, Banpurkar AG, Banat IM, Mone NS, Patil RH, Cameotra SS (2016) Biosurfactant/s from lactobacilli species: properties, challenges and potential biomedical applications. J Basic Microbiol 56(11):1140–1158
162. Elshikh M, Marchant R, Banat IM (2016) Biosurfactants: promising bioactive molecules for oral-related health applications. FEMS Microbiol Lett 363(18):fnw213
163. Elshikh M, Moya-Ramírez I, Moens H, Roelants SLKW, Soetaert W, Marchant R, Banat IM (2017) Rhamnolipids and lactonic sophorolipids: natural antimicrobial surfactants for oral hygiene. J Appl Microbiol 123(5):1111–1123
164. Bouassida M, Fourati N, Krichen F, Zouari R, Ellouz-Chaabouni S, Ghribi D (2017) Potential application of Bacillus subtilis SPB1 lipopeptides in toothpaste formulation. J Adv Res 8 (4):425–433
165. Bujak T, Wasilewski T, Nizioł-Łukaszewska Z (2015) Role of macromolecules in the safety of use of body wash cosmetics. Colloids Surf B: Biointerfaces 135:497–503

166. Stipcevic T, Knight CP, Kippin TE (2013) Stimulation of adult neural stem cells with a novel glycolipid biosurfactant. Acta Neurol Belg 113(4):501–506

167. Ashby RD, Zerkowski JA, Solaiman DK, Liu LS (2011) Biopolymer scaffolds for use in delivering antimicrobial sophorolipids to the acne-causing bacterium Propionibacterium acnes. New Biotechnol 28(1):24–30

168. Meyer-Hoffert U, Zimmermann A, Czapp M, Bartels J, Koblyakova Y, Gläser R et al (2011) Flagellin delivery by Pseudomonas aeruginosa rhamnolipids induces the antimicrobial protein psoriasin in human skin. PLoS One 6(1):e16433

169. Paniagua-Michel Jde J, Olmos-Soto J, Morales-Guerrero ER (2014) Algal and microbial exopolysaccharides: new insights as biosurfactants and bioemulsifiers. Adv Food Nutr Res 73:221–257

Role of Biosurfactants in Agriculture and Soil Reclamation

5

Abstract

The utilization of green agent to attain the goal of sustainable agriculture is the prime necessity. The physiological role of BSs is to decrease the surface tension at the interphase boundaries and therefore permitting microorganisms to metabolize water-immiscible nutrients and in this way, the hydrophobic nutrient is made accessible for uptake and degradation. In agriculture, BSs can be utilized for combating plant pathogens and for increasing the bioavailability of micronutrients for beneficial plant growth-promoting rhizobacteria. Microbial surfactants can broadly be utilized for improving the agricultural lands by soil washing and remediation. The utilization of microbial metabolites to control and prevent spoilage and post-harvest losses is regarded as a substitute for chemical formulations in organic farming, because of their wide applicability in improved crop and food safety. Reclamation of heavy metal polluted soils comprises certain physical approaches such as excavation and dumping of contaminated soil to modern-day landfill locations or microorganisms assisted biological approaches such as the use of BSs or bio-emulsifiers. This book chapter examines the role, potential, and properties of BSs on agricultural practices such as micronutrient solubility, soil washing, and nutrient uptake in plants and their possible applications to resolve environmental contamination problems.

Keywords

Bio-control · Bio-fertilizers · Soil washing · Soil reclamation · Antifungal

5.1 Introduction

In the current world, agricultural production to meet the increasing demands of the human race and environmental sustainability is a matter of huge concern. The utilization of green agent to attain the goal of sustainable agriculture is the prime necessity. Biosurfactants (BSs) are obtained extracellular or cell-associated from a huge population of microorganisms during their growth on hydrophobic nutrients. The physiological role of BSs is to decrease the surface tension at the interphase boundaries and therefore permitting microorganisms to metabolize water-immiscible nutrients and in this way, the hydrophobic nutrient is made accessible for uptake and degradation. Furthermore, BSs chemical structures and surface potential are so diverse, such an emulsifier possibly provides benefits in a particular microbial ecological niche. Various emulsifiers of biological origin have antimicrobial activities [1, 2]. Microbial emulsifiers also play a critical role in maintaining the attachment-detachment of cells to each other and from surfaces. Besides, emulsifiers are having a role in bacterial pathogenesis, cell signaling such as quorum sensing, and biofilm development. Microbial surface-active agents have shown other properties which include cell attachment and it is essential to attain stability under non-favorable conditions; when microbial cells need to find new environments for survival.

In agriculture, BSs can be utilized for combating plant pathogens and for increasing the bioavailability of micronutrients for beneficial plant growth-promoting rhizobacteria [3, 4]. Microbial surfactants can broadly be utilized for improving the agricultural lands by soil washing and remediation. Microbial surfactants can substitute the harsh chemical-based surfactant currently being explored in million-dollar pesticide processing. Therefore, exploring BSs from soil isolates for exploring their potential role in PGPR activities and other linked agricultural applications permits detailed research (Fig. 5.1).

This book chapter examines the role, potential, and properties of BSs on agricultural practices such as micronutrient solubility, soil washing, and nutrient uptake in plants and their possible applications to resolve soil contamination problems.

5.2   Biosurfactants in Agriculture and Environment

5.2.1   Antimicrobial and Antifungal Properties

Various BSs of microbial origin possess antimicrobial potential against a diverse range of plant pathogens and hence they are recognized as effective biocontrol agents for attaining sustainable agriculture practices. BSs obtained from rhizobacteria are considered to have antagonist properties [5–8]. In general, the BSs in agriculture improve the antagonistic potential of microbes and other antimicrobial metabolites products. The glycolipid type of BSs is the most commonly utilized surfactant known to own effective antimicrobial activity (Table 5.1). Additionally, rhamnolipids not showed any adverse impact on environmental and human exposure. The control of the resting spores and mycelia of *Phytophthora sojae* fungal pathogens has been controlled using rhamnolipids obtained from the *Pseudomonas aeruginosa* [28].

Different congeners of the rhamnolipids such as mono-rhamnolipids, di-rhamnolipids, were observed in control of the motility and lysis of zoospores

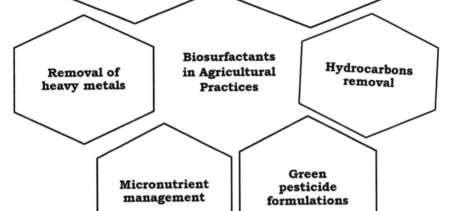

Fig. 5.1 Various roles of biosurfactants in agricultural practices

takes place at the mixture concentration ranging from 8 to 20 mg/L. The mixture of rhamnolipids congeners displayed significant activity not only vegetative fungal mycelia but also extends its antimicrobial activity against the resting zoospores. Similarly, red rot pathogen in sugarcane crop i.e. *Collectotrichium falcatum* was controlled by rhamnolipids analogs Rha-Rha-C10:1 and Rha-C10 [29].

Different analogs of rhamnolipids showed an effective detrimental impact on germinating spores of *Colletotrichum falcatum*. It was demonstrated that disruption of the cell membrane is the key physiological impact on the fungal pathogen. In another case, free diacids obtained from rhamnolipids hydrolyses control the mobility of the *Phytophthora sojae* zoospores at the concentration of 20–125 mg/L [47].

Rhamnolipid was also estimated as a significant agent for the control of Phytophthora capsici, Colletotrichum orbiculare in vitro which is responsible for the brown root rot [33, 34]. The biological control of the plant pathogen, i.e. anthracnose disease was controlled using BSs obtained from Pseudomonas aeruginosa JS29 [48]. The data obtained revealed effective disease using BSs in challenge inoculated spore. Furthermore, the BSs can also efficiently inhibit the pathogen growth in the detached-fruit method in various storage conditions [49–51]. Similarly, the BSs obtained from Pseudomonas sp. GRP3 was found efficient in

Table 5.1 Antifungal and Antimicrobial potential of biosurfactants

S. No.	Biosurfactant	Producing strain	Antifungal/Antibacterial activity	Concentration	References
1	Glycolipid	*P. aeruginosa*	Antifungal activity against *Phytophthora infestans* which causes late blight of potato	100 ppm	Tomar et al. [9]
2	Rhamnolipid	*P. aeruginosa*	Antifungal activity against *M. circinelloides* CCTCC M210113 & *V. dahliae* ATCC 7611 which effect tomato plants by causing wilting	200μg/ml 60μg/ml	Sha and Meng [10]
3	Difficidin	*Bacillus amyloliquefaciens*	Antibacterial activity against *Erwinia amylovora* causing fire blight in apple and pear	1%	Reddy et al. [11]
4	Iturin A	*Bacillus subtilis* KSO3	Antifungal activity against *Gloeosporium gloeosporioides* BA19 which causes anthracnose disease in vegetable and fruits	50μl	Cho et al. [12]
5	Putisolvin like CLPs	*P. Putida* 267	Antifungal activity of *Phytophthora capsici* which causes damping –off cucumber	800μl/20 ml	Kruijt et al. [13]
6	Di-rhamnolipid	*Pseudomonas spp EP-3*	Insecticidal activity against *Myzus persicae* (aphid) causing deformed leaves and spread of plant pathogen and viral diseases	100μl/ml	Kim et al. [14]
7	Sophorolipid	*Rhodotorula babjevae YS3*	Antifungal activity against *Colletotrichum gloeosporioides*, *Fusarium verticillioides*, *Fusarium oxysporum f.sp.pisi*. They cause bitter rot, crown rot, head blight, scab on cereal grains.	62-1000μl/ml	Sen et al. [15]
8	Rhamnolipid	*P. aeruginosa SS14*	Antifungal activity against *Fusarium oxysporum .f.sp.pisi* causing wilting of pea plant	25μg/ml	Borah et al. [16]
9	Bacillomycin D	*B. subtilis AU195*	Antifungal against *Aspergillus flavus* which cause post-harvest rot in crops such as peanuts, cotton seed, corn	3μg	Moyne [17]

10	Surfactin	*B. subtilis*	Antifungal against *Fusarium moniliforme* which contaminates wheat, maize, sorghum, beans, peanuts	50μg/ml	Jiang et al. [18]
11	Fengycin	*B. subtilis FMBJ*	Antifungal against *F.moniliforme* which causes rice bakanae or foot rot disease	2000 ppm	Sarwar et al. [19]
12	Lipopeptides	*B. subtilis GA1*	Antifungal activity against *Botrytis cinerea* causing gray mold disease of apple	25μl	Toure et al. [20]
13	Leu^7-Surfactin	*B. mojavensis*	Antifungal against *F. verticillioides* which cause fungal disease in maize	20μg/ml	Snook et al. [21]
14	Surfactin	*B.siamensis*	Antifungal activity against *Macrophomina phaseolina* goid causing dried root rot,stem rot, charcoal rot in black gram, sorghum, sesame, Sunflower	3%	Hussain and Khan [22]
15	Cyclic lipopeptide	*B. subtilis BBG 131*	Antifungal activity against *Botrytis cinerae* 630 which cause necrosis of grape wine	10μg/ml	Farace et al. [23]
16	Fengycin / Iturin	*B.subtilis* UMAF 6614 UMAF6639 UMAF8561	Antifungal activity against *Podosphaera fusca* causing powdery mildew of cucurbits	1 mg/ml	Romero et al. [24]
17	Fengycin	*B. subtilis B-FS01*	Antifungal activity against *F.moniliforme* causing seedling blight, stalk rot & ear rot	20μg/ml	Hu et al. [25]
18	Poaeamide	*Pseudomonas poae*	Antifungal activity against *Rhizoctonia solani* causing damping off and Rootrot in Sugar beet	50μg ml^{-1}	Zachow et al. [26]
19	Rhammolipid	*P. aeruginosa SS14*	Antifungal activity against *F. verticillioides* causes stalk and ear rot of maize	50μg ml^{-1}	Borah et al. [27]
20	Rhammolipid	*P. aeruginosa*	Antifungal activity against both zoospores and mycelia of *Phytophthora sojae* which causes root and stem rot in soybean	8–20 mg/L 1000 mg/L	Soltani Dashtbozorg et al. [28]

(continued)

Table 5.1 (continued)

S. No.	Biosurfactant	Producing strain	Antifungal/Antibacterial activity	Concentration	References
21	Rha-Rha-C10:1 Rha-C10	P. aeruginosa	Antifungal activity against Colletotrichum falcatum causing red rot disease in sugarcane	40µg/mL	Goswami et al. [29]
22	Rhamnolipids	P. aeruginosa	Antifungal activity against Phytophthora cryptogea which causes root rot disease of witloof chicory	25µg mL^{-1}	De Jonghe et al. [30]
23	Bacillomycin D	B. vallismortis ZZ185	Antifungal activity against F. graminearum and Alternaria alternata which causes scab in wheat	Culture filtrate mL/mL: 0.1,0.5, 1.0 n-butanol extract mg/mL 0.5,1.0,1.5	Zhao et al. [31]
24	Rhamnolipid	P. aeruginosa 47 T2 NCBIM 40044	Antifungal activity against Botrytis cinerea which causes gray rot in grapes dry eye rot on apples	170 mg/L	Haba et al. [32]
25	Rhamnolipid	P. aeruginosa B5	Antifungal activity against zoospore of Cercospora kikuchii which causes leaf spot and blight seed stain in soybeans	50 mg/L	Kim et al. [33, 34]
26	Rhamnolipid	Pseudomonas spp	Antifungal activity against zoospore Phytophthora sojae causing soybean root rot	20 mg/L	Dashtbozorg et al. [35]
27	Rhamnolipid	P. aeruginosa 47 T2 NCBIM 40044	Antifungal activity against Rhizoctonia solani causing eyespot in wheat and other cereals	109 mg/L	Haba et al. [32]
28	Lipopeptide	Bacillus thuringiensis	Prevents pepper anthracnose Colletotrichum gloeosporiodes, Prieris rapae crucivora	100µg ml^{-1}	Il Kim and Chung [36]
29	Kurstakins	B. thuringiensis kurstaki HD-1	Prevents leaf blight of Stachybotrys Charatum	200µg in 80µl and in 200µl	Hathout et al. [37]
30	Iturin A	B. subtilis RB14-C	Controls damping off tomato seedling and cucumber phomopsis root rot R.solani & Phomopsis	3 ml of 10 ppm	Kita et al. [38]

No	Biosurfactant	Organism	Application	Concentration	Reference
31	Surfactin	B. subtilis	Prevents anthracnose disease of trees which results in leaf spots, defoliation, harvest deterioration and mycotoxin production	20–160 mg/L for mycellial germination 0.5 mg/ml for conidial germination	Mohammadipour et al. [39]
32	Iturin A	B.subtilis BS-99-H	Prevents wax apple fruit rot	Conidial &mycellial germination $\geq$0.84 mg/ml	Lin et al. [40]
33	Gageotetrins A-C	B. subtilis	Late blight in cucumber, pepper, tomato, beans	0.02μm	Tareq et al. [41]
34	Rhamnolipid B	Pseudomonas aeruginosa B5	Phytophthora blight and anthracnose	10μg ml^{-1} for zoospore 50μg ml^{-1} for hyphal	Kim et al. [33, 34]
35	Amphisin lipopeptide	Pseudomonas sp. DSS73	Prevents damping off disease in sugar beet	10μgml^{-1}	Andersen et al. [42]
36	Rhammolipid	Pseudomonas aeruginosa PRO1	Root rot disease of witloof chicory	25μgml^{-1}	De Jonghe et al. [30]
37	Massetolid	Pseudomonas fluorescens SS101	Late blight of potato and tomato	100μgml^{-1}	Trans et.al [43]
38	Rhamnolipid	P. aeruginosa	Gray mold disease in grapevine	1 mg ml^{-1}	Varnier et al. [44]
39	Sclerosin	Pseudomonas sp. DF41	Stem rot in canola	300μl	Berry et al. [45]
40	Cyclic lipopeptides	B. amyloliquefaciens ARP23&MEP218	Stem rot in soybean (white mold	100μl &1100μl	Alvarez et al. [46]
41	Bacillomycin D	B. vallismortis zz185	Wheat scab,root disease in wheat	Fusarium Graminearum, Alternaria alternata, Rhizoctonia solani	Zhao et al. [31]

the lysis of plasma membranes of spores of Pythium and Phytophthora fungal pathogens [52]. Hence suggesting rhamnolipids potential role as a biocontrol substance in the direction of control of damping off disease in Chilli and tomato crops.

Other glycolipid BSs such as sophorolipids are also found active against a diverse range of fungi, which comprise plant phytopathogens. Cellobiose lipids were found inhibiting the growth of different phytopathogens such as *Sclerotinia sclerotiorum* and *Phomopsis helianthi* [53]. Moreover, mannosylerythritol lipids are suggested to be used in plant protection because of their antifungal activity.

On the other hand, antifungal activity of Bacillus genus lipopeptides, i.e. Fengycins has also been documented and hence they can be explored in biocontrol of various plant pathogens.

Microbial surfactants are also recognized for their role in accelerating the rate of composting and degradation besides their anti-phytopathogenic properties and so offer extra benefits of utilizing such green surfactants. The BSs which have antimicrobial characteristics against phytopathogens may also distress the other ecological flora of the niche. Therefore, to formulate an effective green surfactant with specificity to control the phytopathogens, the chemical constituents of the BSs may be diverse by varying the production approaches.

5.2.2   Biosurfactants as Biopesticide

The Conventional approach to control the arthropod population is exploring or incorporation broad-spectrum agents and pesticides, which frequently led to a deleterious impact on the environment also. Furthermore, aspects such as the emergence of resistance to the commonly used pesticide between the insect and also rising cost of new synthetic pesticides have to lead to in search of novel eco-friendly arthropod control strategies.

Surfactants are essential as an adjuvant with known antifungal, insecticidal preparations, and herbicides. The chemical surfactant commonly explored in pesticide formulations acts as an emulsifying agent, dispersing and spreading agents, to enhance the effectiveness of pesticides. Besides, these surfactants are utilized in insecticidal preparations in recent agriculture practices (Fig. 5.2) [55]. The lipopeptide of Bacillus genus exhibited the efficient insecticidal potential to control *Drosophila melanogaster* and consequently deliver a promising impression to be utilized as a bioinsecticide [56].

Surfactin obtained from *Bacillus amyloliquefaciens* G1 displayed insecticidal activity against the *Myzus persicae*, green peach aphid by distressing the insect cuticles and persuading substantial dehydration of the membrane to lead to death [57]. BSs derived from the *B. amyloliquefaciens* AG1 inhibit the growth of the *Tuta absoluta* larvae [58]. BSs disrupt the receptors present in the brush-border vesicles of the membrane. The synergistic role of BSs along with the entomopathogenic fungi has a present effective method to control the insects. In a case, rhamnolipids and conidial suspension of the *Cordyceps javanica* and *Beauveria bassiana* were found

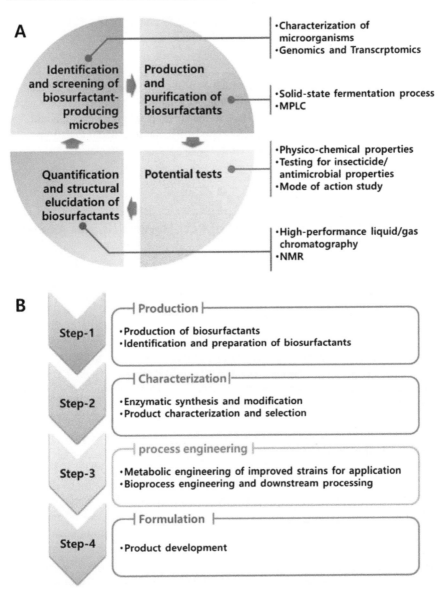

Fig. 5.2 Biosurfactant based novel insecticidal approaches (**a**) Microbial surfactants and application strategies: (**b**) Novel insecticidal production approaches (Courtesy by: [54])

efficient for the control of insects. The results displayed enhanced recovery of the hydrophobic conidia, which ultimately increased their insecticidal potential [59].

BSs have been explored for crop management and protection in integrated pest management strategies, as they are typically degradable by soil microorganisms and ecologically friendly. BSs derived from the Bacillus strain displayed insecticidal activity against the *Myzus persicae.* The lipopeptide containing leucine displayed excellent insecticidal potential than the isomers containing valine. BSs displayed a potential for utilization as biocontrol metabolites in an integrated pest control strategy ([60]).

Agriculture associated formulations such as pesticides produced with the help of BSs can be extensively utilized in agricultural practices. The requirement for agro-chemical preparations is to prepare effective formulation strategies and to attain this goal; various industries can explore a mixture of BSs in diverse combinations with the bio-polymers to make exceptional formulations for sustainable agricultural practice.

5.3   Biosurfactants in Control of Post-Harvest Disease Control

The utilization of microbial metabolites to control and prevent spoilage and post-harvest losses is regarded as a substitute for chemical formulations in organic farming, because of their wide applicability in improved crop and food safety. The BSs based interventions have the activity as a biological solution to control fungal disease, which could be capable of biocontrol molecule for attaining organic agriculture practices.

Postharvest spoilage has levied serious problems for the storing of various fresh fruits and vegetables produce, particularly acidic fruit and succulent vegetables that frequently agonize from rotting diseases [61]. It is estimated that, 40% of total agricultural produce is lost due to the postharvest losses [62]. Various fungal pathogens are associated with the postharvest losses during the transportation and storage such as *A. alternata, Fusarium semitectum, Rhizopus stolonifer*, and *Trichothecium roseum* [63, 64].

The chemical based fungicide are mainly used to control the postharvest losses [65]. But, various drawbacks have been associated with the utilization of chemical-based fungicides such as resistance towards the fungicides, deleterious effects on human and animal health, and environmental concerns [66, 67]. Such impacts on human and environmental health concerns, hence there is a requirement to find substitutes to chemical fungicides. Biological prevention of postharvest losses has appeared as one of the actual non-chemical substitutes [68], and serves as outstanding biological control agents. Lipopeptide BSs such as surfactin, iturin, and fengycin are known for the antibacterial and antifungal potential. The lipopeptide BSs displayed a non-specific mechanism of cell death which anticipated that they will be potent in controlling an extensive range of phytopathogens [68–74].

Biosurfactants of the *Bacillus subtilis* were found effective in the control of postharvest diseases of melon [75]. The growth of the pathogen, i.e. *Alternaria alternata* was reduced by 77.2% after co-incubation with *B. subtlis.* The nutritive values of the treated fruits were found and restored as the concentration of ascorbic acid increases, whereas the amount of the organic acid accumulated was minimal, and fruit sustains the water content and turgidity at ambient temperature. BSs are not only protecting the fruits from postharvest diseases but also provide physiological benefits during the storage.

The lipopeptides BSs have been observed as efficient antifungal agents [76]. Rhamnolipids were also found to have efficient antifungal properties. In a case, the antifungal potential of the rhamnolipids has been assessed against *Alternaria alternata* and the underlying possible mode of action was investigated. The decay occurrences of cherry tomatoes by *A. alternata* coated with rhamnolipids were expressively decreased. Rhamnonolipid was found effective against spores and vegetative mycelial growth on solid and submerged surfaces. Besides, the synergistic efficacy of rhamnolipids and essential oil reduced the total amount of fungicide required. Furthermore, the morphological and structural abrasions were observed with electron microscopy which revealed major alterations in the hyphae.

Antifungal BSs derived from the *Bacillus subtilis* were found effective in the control of the fruit rot diseases in tomato caused by the *Botrytis cinerea.* Coating of the tomato fruits with bacillus BSs reduced the postharvest losses by 79%. *Bacillus subtilis* was also found to produce a high amount of chitosanase and which might be appreciated for the various industrial and agricultural roles.

A combination of BSs and different biocontrol microorganisms also presents an effective method to control postharvest diseases. A mixture of rhamnolipids (500 ug/mL) and biocontrol yeast, i.e. *Rhodotorula glutinis* has been reported for the control of the cherry tomato caused by *Alternaria alternata* [77]. Moreover, amalgamation of *R. glutinis* with rhamnolipid primarily encouraged peroxidase, polyphenoloxidase, and phenylalanine ammonia lyase activities of cherry tomato, which were stronger than that of a solo treatment.

Although BSs are considered as a type of surfactant, BSs cannot develop a homogenous and stable formulation due to their partial solubility in aqueous suspensions. To make BSs advantageous to industrial applications, selecting some suitable solvents to mix BSs into a homogenous formulation is very significant. There is still a scope to perform trails to known the interaction of BSs with fungal pathogens.

5.4 Soil Health and Micronutrients Availability

Micronutrient deficiencies in the soil are a major problem worldwide that affects gross agricultural production and human health [49, 51, 78]. The projected increase in the global population needs high agricultural productivity to increase pointedly within the next few years [79]. Soil systems have gradually more deficient in a diverse range of micronutrients as soil systems are exhausted due to continuous crop cultivation.

Soil systems around the world such as India, China, and Australia are suffering from essential micronutrient deficiency problems due to high crop cultivation demands [80]. The concentration of micronutrients available in the agricultural produce signifies the nutrient concentration in soils and also the nutrient availability of soil system [81]. Inadequate availability of micronutrients in food will be

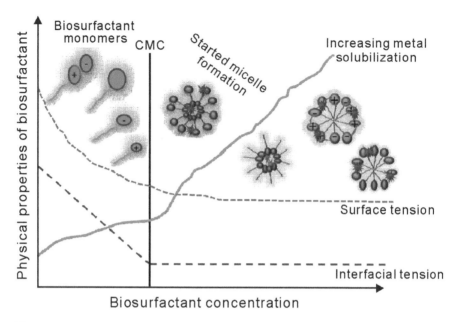

Fig. 5.3 Interrelationship between BSs concentration and metal solubility [85]

responsible for malnutrition in developing countries [82]. Micronutrient deficiency in soil systems is not just because of the constant agricultural practices but inadequate use of organic matter such as compost and other agricultural leftovers. Even with the continuous utilization of chemical-based fertilizers, erosion of the top organic layer of the soil results in the exhaustion of the physical, chemical, and biological properties of agricultural soils [83].

BSs are multifunctional amphiphiles and have been documented as green surfactants due to their biodegradability and being eco-friendly compositions [84]. BSs can enhance the removal and uptake of divalent metal bioavailability in agricultural and industrially affected soils (Fig. 5.3) [86]. There is an interrelationship between BSs concentration and the physical properties of the soil system. At the CMC concentration of BSs molecules, the solubility of the heavy metals present in the soil system increases [85]. The microbial surfactants are known to enhance the micronutrient bioavailability, improved wettability, nutrient uptake, and recycling in soil systems, and also achieve a better uniform diffusion of nutrients mixtures [49, 51].

The microbial surfactants can be an attractive strategy in comparison to the chemical-based surfactants, especially in micronutrient scarcity soil systems [87–89]. Henceforth, there is a requirement to categorize and evaluate chelators/solubilizers with possibly effective micronutrient-solubilizing potential and environmentally friendly, i.e. biosurfactants. Biosurfactants play an important role in micronutrient chelation, the process of adsorption and desorption, and removal of the heavy metals from the soil surface interfaces resulting an enhanced micronutrient availability in soil systems [90].

Microbial surfactants can enhance the flow of metals by reducing the interfacial tension by micelle development between soil and metal ions [89]. BSs can absorb the metal ions by reducing the interfacial tensions and transfer them to the root rhizosphere [91–94]. In a report, Wang and Mulligan [95] demonstrate that the incorporation of a rhamnolipid enhanced the dissolvability of the metal which ultimately increase the mobility in the soil. BSs can enhance the availability of metals, such as zinc, manganese, copper, and iron in the soil to root rhizosphere in various ways.

The strategy for improving the absorption of micronutrients in a plant system by the use of BSs to the root rhizosphere might be a significant approach in generating sustainable agriculture practices [96]. BSs obtained from the Bacillus sp. strain J119 were reported for their potential to enhance the plant growth and absorption of trace metals in the different crops of canola, maize, sudangrass, and tomato Sheng et al. [97]. Furthermore, Stacey et al. [98] observed the effect of the monorhamnosyl and dirhamnosyl rhamnolipids complexes with copper and zinc, improving the absorption of such metals by *Brassica napus* and *Triticum durum* crops via the root system.

5.5 Environmental Applications

5.5.1 Degradation of Polycyclic Aromatic Hydrocarbon (PAH)

The PAHs are major carcinogenic contaminants that are resilient to degradation in soil because of their hydrophobic behavior. The adverse impact of the PAHs on human, animal, and soil health has resulted in widespread reports on the removal of soils polluted with PAHs [99]. The unique characteristics of such toxic substances are extremely hydrophobic in nature. The PAHs contaminants adsorb onto the organic stock of solid particles, resulting in obstinate micropollutants in the soil system. The PAH accumulation is due to anthropogenic practices, such as the burning of agricultural biomass, insufficient combustion of fossil fuel, industrial practices, and oil spills [100, 101]. Besides this, PAHs contamination in the soil also results from various other activities such as soils from coal storage sectors, gas manufacturing plants, and spillage of coal tar leads to the PAHs contamination. As the ring structure in rises, the substances become more resilient to degradation due to volatility, solubility behavior, and increase in sorption.

PAHs are of huge concern due to well-recognized genotoxic, mutagenic, and cancer-causing effects [102]. PAHs are furthermore metabolized degraded in ways comparable to single ring aromatics in the meantime PAHs are demolished one ring at a given time. Different treatment strategies including organic, physicochemical, and as well thermal processes have been developed for the reclamation of polluted sites [91].

Microbial degradation can be limited by the availability of soil-associated PAHs due to the low aqueous dissolvability, hydrophobic nature, which is exaggerated by the extreme accumulation of foreign compounds in field polluted soils [103]. Microbial surfactants are well-recognized to enhance the availability of such hydrophobic substances, hence it aids to daze the diffusion-related mass transfer restrictions [104].

BSs derived from the microorganisms are considered as an effective microbial strategy that impacts the availability of hydrophobic substances by altering the cell surface hydrophobicity or by mixing and emulsifying such hydrophobic substances [93].

The lipopeptide BSs derived from the *Bacillus cereus* SPL-4 were utilized for the removal of the PAH on a laboratory-scale trial [105]. In the present experiments, the removal of PAHS has been enhanced effectively by the high molecular weight BS treatment as compared to the surfactant-free controls. The degradation of 5 and 6 ring PAHs was efficiently improved ($p < 0.05$) in the elevated surfactant concentration as compared to the low concentrations. The outcome of the study suggests that BSs assisted removal by bacterial consortium may be an effective practical bioremediation approach for long-accumulated PAH-polluting soils.

Xia et al. [106] observed that sludge washing processing involving lipopeptides obtained from *Pseudomonas sp.* WJ6 has significant removal efficiency i.e. 92.46%. A microbial strain of Pseudomonas sp. was isolated from Nigerian soil has been identified as a lipopeptide producer based on its potential to degrade pyrene and diesel oil [107]. A different strain of the actinomycete such as *Streptomyces* spp. has been isolated from Algeria plain has the potential to metabolize petroleum and naphthalene. Based on the liquid and gas chromatography analysis, it was found that 82.36% naphthalene was degraded after the 12 days of the incubation period [108]. The BSs derived from the *Pseudomonas aeruginosa* utilizing coking wastewater as a nutrient source was found efficient to remedaite the contaminated soil [109]. The BSs were found stable at a different pH range such as 3.5–9.5 at a CMC of 96.5 mg/L and 0–15% salinity into production media consisting of cooking wastewater. Microbial strain obtained from the extreme hot spring with an efficient potential for PAHs degradation and BSs stability at an extreme temperature provides a novel advantage for the removal of PAHs. Mehetre et al. [110] isolated the thermophilic lipopeptide producer, i.e. Aeribacillus able to degrade a mixture of PAHs. The high rate of degradation was achieved at a temperature of 50 °C [110]. Enhanced degradation of the naphthalene was recorded with the rhamnolipids derived from the *Pseudomonas* sp. The produced rhamnolipid enhanced swift degradation of naphthalene when compared to the depletion without rhamnolipid.

Sluggish diffusion in soil aggregates, strong bonding between organic pollutant and soil system will be responsible for the non-availability. In a given situation, the availability hindrance of the PAHs is radiated through BS supplementation. Such results are particularly significant for high molecular weight PAHs and maybe a real benefit for the degradation of accumulated hydrocarbon in polluted soil sites.

Therefore, PAHs degradation by BSs supplementation displays the effective use for the degradation at polluted environmental sites.

5.5.2 Heavy Metal Removal and Soil Washing

In the present industrial scenario, heavy metals contamination of soil and agricultural land are becoming part of the grave environmental concern. Pollution of soil systems with heavy metals is tremendously hazardous to animal and human health within the ecosystem. Even the very low concentration accumulation of heavy metals poses a serious threat within the agricultural soils with various detrimental outcomes. Common examples of heavy metals include zinc, nickel, cadmium, lead, chromium, arsenic, and copper, etc. [111]. At present, there are various approaches explored to clean up soils polluted with heavy metals. Reclamation of heavy metal polluted soils comprises certain physical approaches such as excavation and dumping of contaminated soil to modern-day landfill locations or microorganisms assisted biological approaches such as the use of BSs or bio-emulsifiers. Different synthetic surfactants had been utilized to clean heavy metal polluted soils. The synthetic surfactants are considered to be toxic compounds and may lead to other environmental concerns because of their shelf life or degradability in the soil systems [112].

In evaluation with synthetic surfactants, microbial surfactants obtained from plants and microbial cells have display improved performance well-thought-out appropriate in heavy metal removal from contaminated soil [3, 113, 114] (Figs. 5.4 and 5.5).

Soil washing mentions the removal of heavy metals by utilizing various reagents and solutions [116, 117] which remove metal from the soil. Presently, the utilization of appropriate substances for bio-leaching of heavy metals from polluted soils has been established as a substitute for various conventional methods for the cleanup of polluted soils.

The performance of BSs for removal of heavy metal polluted soil is based on their potential to form complexes with metals. The anionic BSs form complexes with accumulating metals in a non-ionic complex by forming ionic bonds. Such kind of

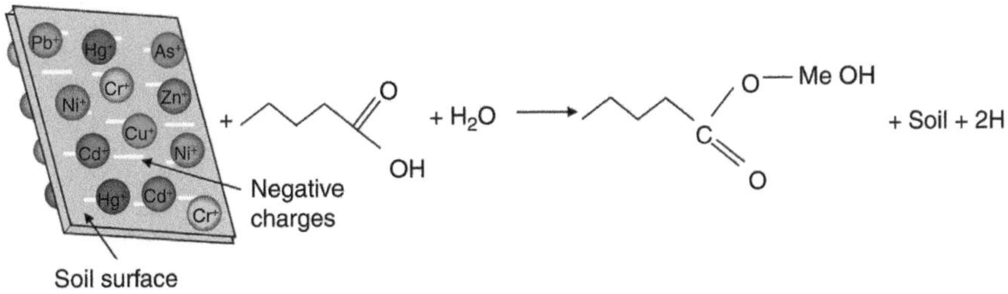

Fig. 5.4 Mechanism of removal of the heavy metals using BSs from contaminated soil and water sample (Courtesy by: [115])

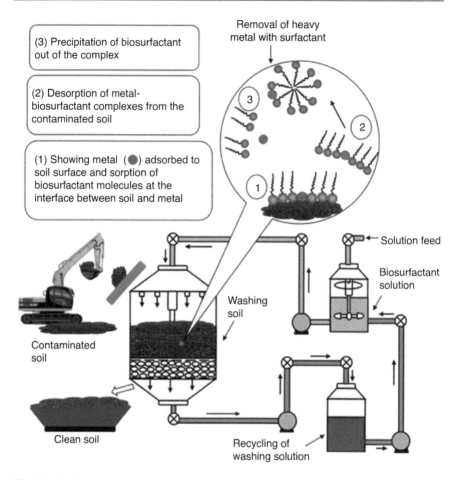

Fig. 5.5 Possible technology flowchart for the removal of the heavy metal removal from polluted soil using BSs (Courtesy by: [115])

complex formations is much stronger than the bond formation between soil and metal. The presence of the BSs in the soil system reduces the interfacial and surface tension. Heavy metal ions can form complex with the BS due to the formation of the micelles. The fatty acids polar head arrangements in a micellar structure bind and mobilize the metals present in their solubilized form.

The mode of heavy metal remediation from affected soil utilizing ionic BSs is depicted in Fig. 5.4. In general, various steps are involved in the elimination of heavy metals exploring the possibilities of soil washing utilizing BSs solutions. The BSs reduce the interfacial tension between heavy metals and soil particles which leads to the formation of wet soil and metal comes out in an aqueous solution. At that point, the metal ion is absorbed by BSs and entrapped in the micelle through electrostatic bonding. Lastly, the BSs can be recovered using the membrane separation method.

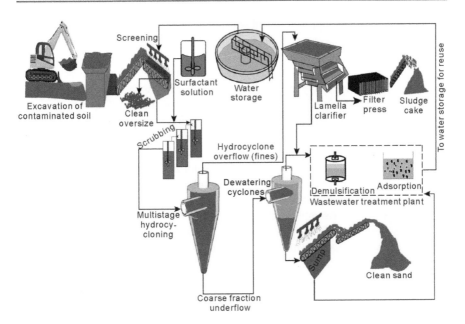

Fig. 5.6 Diagram of soil washing by ex-situ soil approach (Courtesy by: Alternative Remedial Technologies, Inc. [123])

Anionic BSs derived from the *Candida sphaerica* utilized in the removal of the zinc, iron, and led removal using soil washing and it was found to remove removal of the metals in the range of 65–80% [113]. Rhamnolipids are also involved in the removal of copper, zinc, and lead using soil washing [118]. In another case, Santos et al. [119] challenged the soil sample with Cu, Pb, and Zn, and soil was washed with BSs obtained from the *C. sphaerica* UCP0995. BS based washing removed significant metal content (6–35%) of the challenged concentration. The BSs obtained from the *Acidithiobacillus thiooxidans* and *Acidithiobacillus ferrooxidans* jointly removed 70% zinc while rhamnolipids removed only 52% Zinc when used alone [120]. In a report, 52% of the metal has been removed by bioleaching using glycolipids derived from Burkholderia sp. Z-90 [121]. The rhamnolipids obtained from the *P. aeruginosa* eliminated 98.83% and soil clay of zinc from feldspar soil and soil clay respectively [122].

In the process of the soil washing, the contaminated soil is excavated out and BSs solution is mixed with an appropriate extractant solution depending upon the type of metal and soil (Fig. 5.5). The leachate obtained from the previous study using metal precipitation, ions exchange, metal chelation, or metal adsorption is transferred from soil to liquid phase, and subsequently, metal was obtained from the metal leachate (Fig. 5.6) [124]. After treatment, the treated soil that fulfills regulatory measures is often recharged to the original site.

Additionally, soil washing is a strategy that is a fast approach that may meet explicit criteria with no long-term responsibility [117]. Because of its significant efficiency, the present approach of soil washing is taken into consideration as one of the effective cost-effective technologies for soil remediation [125].

Although EDTA is most commonly utilized compound in the soil washing approach [126]. But BS like rhamnolipids is the most investigated surfactants for the removal of the heavy metals at different pH range and concentrations [127]. Along with rhamnolipid lipopeptides surfactants at their CMC concentration are an efficient and environmentally amicable strategy for the removal of heavy metals from contaminated soil. Surfactin and Fengycin BSs obtained from the *Bacillus subtilis* removed a huge amount of heavy metals such as Cd (44.2%), Co (35.4%), Pb (40.3%), Ni (32.2%), Cu (26.2%), and Zn (32.07%)[128].

Rhamnolipids and di-rhamnolipids were involved in the removal of Cd and Pb from contaminated soil after 36 h of treatment. This displayed that di-rhamnolipid extensively helps to assemble cadmium and lead ions in a capacity of Cd > Pb [129]. BSs derived from the *Lactobacillus pentosus* cultivated in hemicellulosic sugars hydrolysate obtained from vine trimming sugar were explored in the removal of soil polluted with octane.

BSs obtained from the LAB have the potential to remediate the octane in the soil system. It was observed that BSs obtained from the *L. pentosus* enhanced the biodegradation of octane in the soil system in the 15 days of incubation. It was measured as the concentration of degradative octane varies in the range of 58.6–62.8%, for soil comprising about 700 and 70,000 mg/kg of octane. Under controlled conditions, in absence of the BSs, the rate of degradation of the octane was only approximately 1.2 and 24%.

Artificially spiked soil samples with cadmium and lead in a batch system have been performed using sophorolipids derived from *Starmerella bombicola*. The removal performance of sophorolipids, synthetic surfactants, and distilled water has been compared. Furthermore, approximately 83.6% and 44.8% of Cd and Pb were removed by 8% of crude acidic sophorolipids. In the present study, acidic SLs displayed appropriate aqueous solubility as compared to the lactonic sophorolipids in enhancing the removal of heavy metal polluted soils. The cell-free supernatant containing sophorolipids obtained from *S. bombicola* in its partial purification and attained approximately 95% of cadmium and 52% of lead.

So, to make microbial surfactant enhanced technologies such as soil washing added economically practicable, additional work on BSs recovery and recycling is also obligatory. In comparison, there is no appropriate single soil pollution removal approach; the combination of two or more such approaches may be essential for the effective remediation of soils, type, and nature of the contaminant, the ecological parameter, and the functional conditions.

5.5.3   Microbial Enhanced Oil Recovery

Microbial enhanced oil recovery (MEOR) using microbial surfactants is a type of strategy to get the improved oil recovery method, that kind of methods are only

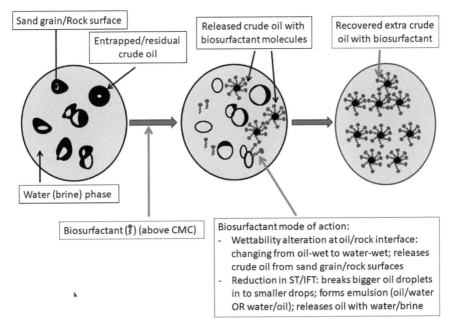

Fig. 5.7 Different mechanisms of BSs mediated EOR (Courtesy by: [137])

explored in a condition where the traditional techniques such as excavation, pressurized extraction, and pumping out of oil from oil reservoirs. Surface-active agents are amphiphilic molecules, constituting both hydrophilic and hydrophobic moieties. Synthetic surfactants are one of the extensively explored agents in various sectors such as household applications, cleaning, cosmetics, and petroleum [130]. Surfactants have been classified based on their structural compositions and charge as anionic, cationic, and non-ionic [131].

Surfactants are the most expensive and one of the favored compounds for EOR applications. Certain issues such as adsorption to soil particles, recalcitrant nature, and thus not so ecologically friendly [132]. On the other hand, BSs have various advantages as compared to chemical surfactants like production using renewable substrates, low CMC potency under an extensive range of extreme environmental conditions, and low or non-toxic [133, 134].

BSs assisted EOR approaches can be utilized in two probable ways such as in-situ and ex-situ type. In the in-situ EOR approach, potential BS producer strain or nutrients to augment the growth are injected into the oil reservoirs or drilled well, tailed by shut-in phase with water flooding consequently. Different low-cost conventional feedstocks have been used as nutrients such as sugarcane molasses, sugar syrups, inorganic nitrogen source as nitrate salts [93, 135]. In the ex-situ EOR approach, the microbial metabolites are derived from the BS producers outside the oil well and injected straight into the oil wells [136]. BSs are very effective in

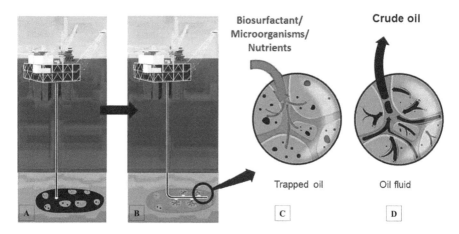

Crude oil

Biosurfactant/
Microorganisms/
Nutrients

Trapped oil   Oil fluid

Fig. 5.8 Process of EOR of crude oil using microbial surfactants. (A) Oil recovery utilizing the natural pressure of the oil well. (B) Oil reservoir pressure reduced. (C) Chief approach of BS utilized for the oil release. (D) Oil well pressure reestablished enabling oil extraction (Courtesy by: [137])

enhancing the EOR from low lying oil reservoirs by various mechanisms (Figs. 5.7 and 5.8).

Different approaches for the efficient potential of the BSs producing strain are evaluated in EOR utilizing the sand-pack columns under oil well conditions are trailed by primary screening and pick out the potential strains. The small-scale laboratory EOR set-ups are demonstrated as sand-pack column or core-flood (exploring various sandstone, carbonate, dolomite core plugs [137]. The sand-pack models are utilized only in primary experiments or in-situ (Fig. 5.9), but core-flood set-ups are favored to simultaneously confirm EOR potential at the field trial or ex-situ methods (Fig. 5.10).

However, in ex-situ EOR applications, BSs are predominantly obtained just outside the site and are then injected into oil well [137]. The BSs obtained from the *Bacillus licheniformis* have been investigated for the EOR efficacies and 25.78% oil recovery was attained [139]. In another report, Zou et al. [140] observed the potential obtained from *Acinetobacter baylyi*. It showed the potential to reduce the surface tension of water 35 mN/m from 65 mN/m, similarly, the interfacial tension reduces to 15 mN/m from 15 mN/m. The potential of EOR observed that 28% more oil recovery from the leftover oil subsequent water flooding.

BS-EOR can be effortlessly carried out ex-situ, as there is no requirement of the process modification of the existing water injection method. Different process set-ups were reported to the study the BS-EOR approach. Rhamnolipids obtained from the *Pseudomonas aeruginosa* were used with glass-bead packs and cores of sandstone [141].

The shake flask and pilot-scale EOR set-ups displayed 10–85% enhanced oil recovery, but the field trial data displayed approximately 3–10% enhanced oil recovery, which is still inspiring. Several laboratory experimental set-ups and reports

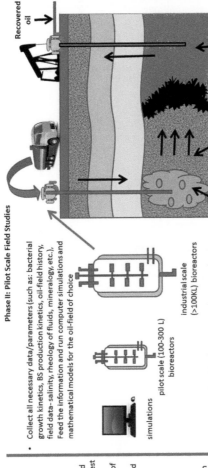

Phase I: Laboratory Studies

- Isolation of exogenous/identification of indigenous biosurfactant (BS) producing microbes

- Screening /identification reported BS producing microbes using latest molecular biology tools

- Initial lab experiments for effect of different nutrient addition on BS production in liquid medium/solid substrates (core-plugs), under anaerobic conditions, and environmental conditions at oil-field of interest

- Analyze different microbial products produced (such as BS, VFA, gases, solvents, etc.), growth patterns (propagation in rocks-time taken to populate in core-plugs), and effect on EOR using glass micro-models, sand-pack columns and/or core-flood experiments

Phase II: Pilot Scale Field Studies

- Collect all necessary data/parameters (such as: bacterial growth kinetics, BS production kinetics, oil-field history, field data - salinity, rheology of fluids, mineralogy, etc.), Feed the information and run computer simulations and mathematical models for the oil-field of choice

simulations   pilot scale (100-300 L) bioreactors

industrial scale (>100KL) bioreactors

- For pilot scale oil-field studies grow bacteria at large Scale in pilot scale (100-300 L) OR industrial scale (>100KL) bioreactors – based on requirement/scale of applications

- Injection of selected cheaper nutrients to the oil wells OR exogenous BS producing facultative or anaerobic bacteria along with selected nutrients to the oil wells: Single well application OR Multipoint application

Recovered oil

Injected BS producing microbes   In-situ produced BS   Oil front   Oil bearing zone

- After shut-in phase flood with formation brine and analyze for %EOR, propagation of injected bacteria/indigenous bacteria, produced BS, etc.

Phase III: Full-Field Application(s)

Based on the results obtained from pilot studies, decide to go ahead for full-field applications or not

Fig. 5.9 Stepwise "Laboratory-to-field trial" in-situ BS mediated EOR (Courtesy by: [137])

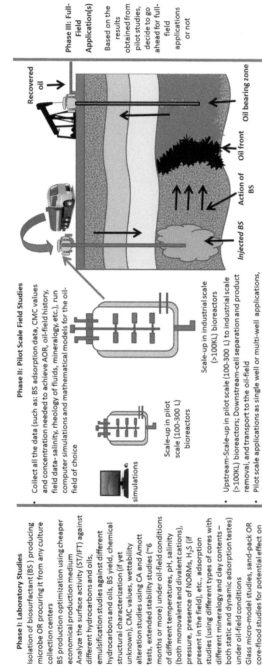

Fig. 5.10 Graphical step wise "Lab-to-field" use of ex-situ approach involving the BSs mediated EOR steps (Courtesy by: [137])

Phase I: Laboratory Studies

- Isolation of biosurfactant (BS) producing microbe OR procuring it from any culture collection centers
- BS production optimization using cheaper economical production medium
- Analyze the surface activity (ST/IFT) against different hydrocarbons and oils, emulsification studies against different hydrocarbons and oils, BS yield, chemical structural characterization (if yet unknown), CMC values, wettability alteration studies using CA and Amott tests, extended stability studies (~6 months or more) under oil-field conditions of different temperatures, pH, salinity (both monovalent and divalent cations), pressure, presence of NORMs, H₂S (if present in the field), etc., adsorption studies (using different types of cores with different mineralogy and clay contents – both static and dynamic adsorption testes) under oil-field conditions
- Glass micro-model studies, sand-pack OR core-flood studies for potential effect on EOR using fluids (formation water and crude oil) and rocks (core-plugs) from the oil-field of interest, at different concentrations (above and below CMCs)

Phase II: Pilot Scale Field Studies

- Collect all the data (such as: BS adsorption data, CMC values and concentration needed to achieve AOR, oil-field history, field data-salinity, rheology of fluids, mineralogy, etc.), run computer simulations and mathematical models for the oil-field of choice

simulations

Scale-up in pilot scale (100–300 L) bioreactors

Scale-up in industrial scale (>100KL) bioreactors

- Upstream-Scale-up in pilot scale (100–300 L) to industrial scale (>100KL) bioreactors; Downstream-cell separation and product removal, and transport to the oil-field
- Pilot scale applications as single well or multi-well applications, either as slug(s) of biosurfactant solutions (concentrated or diluted, based on the mineralogy of the field – sandstone or carbonate etc.) or dissolving the biosurfactant in injection water and directly injecting in the wells

Phase III: Full-Field Application(s)

Based on the results obtained from pilot studies, decide to go ahead for full-field applications or not

Recovered oil

Injected BS   Action of BS   Oil front   Oil bearing zone

- longer shut-in phase not required, collect fluid data from producer well(s) and analyze %EOR, outbreak of BS from the producing wells displayed the ease of application, stability on various environmental conditions, robust, ST/IFT decrease, emulsification potential, exceptional EOR recovery rate. While both in-situ and ex-situ BS mediated EOR methods have huge potential and drawbacks at the same time, the published reports and data specify that currently, in-situ BS-EOR is of ideal industrial selection.

5.5.4   Role in Waste Treatment

Microbial surfactants are amphipathic agents composed of hydrophilic and hydrophobic components, such characteristic confers the opportunity of various applications in waste management and treatments.

BSs has been useful for the removal of the insoluble paint waste from the wastewater [142]. The BS obtained from the *Corynebacterium aquaticum* and *Corynebacterium* spp. was utilized to clean industrial waste. The BSs that displayed reduced surface tension and significant emulsifying potential were utilized in the removal of insoluble paint. The fish and baggase waste have been explored as a conventional nutrient source for the production of BSs using *Corynebacterium aquaticum* displayed appropriate BS production.

The emulsification activity was approximately 87.6 % with a reduction of surface tension to 33.9 mN/m. The utilized BSs displayed potential utilization in-solubilization and removal of the paint. Therefore, the residues of fish and sugarcane bagasse displayed appropriateness as nutrient sources to produce BSs. Besides, the key application in preliminary paint removal provided great results that can be utilized in the future.

5.6   Future Directions and Conclusions

The utilization of greener BSs agent to achieve the goal of sustainable environmental and agricultural solutions is the prime concern of the organizations. The physiological role of BSs is to decrease the surface tension at the interphase boundaries and therefore permitting microorganisms to metabolize water-immiscible nutrients and in this way, the hydrophobic nutrient is made accessible for uptake and degradation. In agriculture, BSs can be utilized for combating plant pathogens and for increasing the bioavailability of micronutrients for beneficial plant growth-promoting rhizobacteria. But still, various challenges have been associated such as purity of the BSs required, cost of the production technologies, and interactions of soil and environmental matter with the BSs. Future research should be in the direction to attain the possible role in agriculture and environmental cleanups. Mass productions of the partially purified BSs will attain the solution and functional synergies with existing technologies will enhance the role of biosurfactants in near future.

References

1. Ron EZ, Rosenberg E (2001) Natural roles of biosurfactants: minireview. Environ Microbiol 3 (4):229–236
2. Kumar P, Sharma PK, Sharma PK, Sharma D (2015) Micro-algal lipids: a potential source of biodiesel. J Innov Pharm Biol Sci 2(2):135–143
3. Sachdev DP, Cameotra SS (2013) Biosurfactants in agriculture. Appl Microbiol Biotechnol 97 (3):1005–1016
4. Sharma D, Dhanjal DS, Mittal B (2017) Development of edible biofilm containing cinnamon to control food-borne pathogen. J Appl Pharm Sci 7(01):160–164
5. Nihorimbere V, Ongena M, Smargiassi M, Thonart P (2011) Beneficial effect of the rhizosphere microbial community for plant growth and health. Biotechnol Agron Soc Environ 15 (2):327–337
6. Singh A, Van Hamme JD, Ward OP (2007) Surfactants in microbiology and biotechnology: part 2. Application aspects. Biotechnol Adv 25(1):99–121
7. Sharma D, Saharan BS (eds) (2018) Microbial cell factories. CRC Press, Boca Raton
8. Singh P, Patil Y, Rale V (2019) Biosurfactant production: emerging trends and promising strategies. J Appl Microbiol 126(1):2–13
9. Tomar S, Singh BA, Khan MA, Kumar S, Sharma S, Lal M (2013) Identification of Pseudomonas aeruginosa strain producing biosurfactant with antifungal activity against Phytophthora infestans. Potato J 40(2)
10. Sha R, Meng Q (2016) Antifungal activity of rhamnolipids against dimorphic fungi. J Gen Appl Microbiol 62(5):233–239
11. Reddy AS, Chen CY, Baker SC, Chen CC, Jean JS, Fan CW et al (2009) Synthesis of silver nanoparticles using surfactin: a biosurfactant as stabilizing agent. Mater Lett 63 (15):1227–1230
12. Cho SJ, Hong SY, Kim JY, Park SR, Kim MK, Lim WJ et al (2003) Endophytic Bacillus sp. CY22 from a balloon flower (Platycodon grandiflorum) produces surfactin isoforms. J Microbiol Biotechnol 13(6):859–865
13. Kruijt M, Tran H, Raaijmakers JM (2009) Functional, genetic and chemical characterization of biosurfactants produced by plant growth-promoting Pseudomonas putida 267. J Appl Microbiol 107(2):546–556
14. Kim SK, Kim YC, Lee S, Kim JC, Yun MY, Kim IS (2011) Insecticidal activity of rhamnolipid isolated from Pseudomonas sp. EP-3 against green peach aphid (Myzus persicae). J Agric Food Chem 59(3):934–938
15. Sen S, Borah SN, Bora A, Deka S (2017) Production, characterization, and antifungal activity of a biosurfactant produced by Rhodotorula babjevae YS3. Microb Cell Factories 16(1):1–14
16. Borah SN, Goswami D, Lahkar J, Sarma HK, Khan MR, Deka S (2015) Rhamnolipid produced by Pseudomonas aeruginosa SS14 causes complete suppression of wilt by Fusarium oxysporum f. sp. pisi in Pisum sativum. BioControl 60(3):375–385
17. Moyne AL, Shelby R, Cleveland TE, Tuzun S (2001) Bacillomycin D: an iturin with antifungal activity against Aspergillus flavus. J Appl Microbiol 90(4):622–629
18. Jiang J, Gao L, Bie X, Lu Z, Liu H, Zhang C et al (2016) Identification of novel surfactin derivatives from NRPS modification of Bacillus subtilis and its antifungal activity against Fusarium moniliforme. BMC Microbiol 16(1):31
19. Sarwar A, Hassan MN, Imran M, Iqbal M, Majeed S, Brader G et al (2018) Biocontrol activity of surfactin a purified from Bacillus NH-100 and NH-217 against rice bakanae disease. Microbiol Res 209:1–13
20. Toure Y, Ongena MARC, Jacques P, Guiro A, Thonart P (2004) Role of lipopeptides produced by Bacillus subtilis GA1 in the reduction of grey mould disease caused by Botrytis cinerea on apple. J Appl Microbiol 96(5):1151–1160

21. Snook ME, Mitchell T, Hinton DM, Bacon CW (2009) Isolation and characterization of Leu7-surfactin from the endophytic bacterium Bacillus mojavensis RRC 101, a biocontrol agent for Fusarium verticillioides. J Agric Food Chem 57(10):4287–4292

22. Hussain T, Khan AA (2020) Bacillus subtilis HussainT-AMU and its antifungal activity against potato black scurf caused by Rhizoctonia solani on seed tubers. Biocatal Agric Biotechnol 23:101443

23. Farace G, Fernandez O, Jacquens L, Coutte F, Krier F, Jacques P et al (2015) Cyclic lipopeptides from B acillus subtilis activate distinct patterns of defence responses in grapevine. Mol Plant Pathol 16(2):177–187

24. Romero D, de Vicente A, Rakotoaly RH, Dufour SE, Veening JW, Arrebola E et al (2007) The iturin and fengycin families of lipopeptides are key factors in antagonism of Bacillus subtilis toward Podosphaera fusca. Mol Plant-Microbe Interact 20(4):430–440

25. Hu LB, Shi ZQ, Zhang T, Yang ZM (2007) Fengycin antibiotics isolated from B-FS01 culture inhibit the growth of Fusarium moniliforme Sheldon ATCC 38932. FEMS Microbiol Lett 272 (1):91–98

26. Zachow C, Jahanshah G, de Bruijn I, Song C, Ianni F, Pataj Z et al (2015) The novel lipopeptide poaeamide of the endophyte Pseudomonas poae RE* 1-1-14 is involved in pathogen suppression and root colonization. Mol Plant-Microbe Interact 28(7):800–810

27. Borah SN, Goswami D, Sarma HK, Cameotra SS, Deka S (2016) Rhamnolipid biosurfactant against Fusarium verticillioides to control stalk and ear rot disease of maize. Front Microbiol 7:1505

28. Soltani Dashtbozorg S, Kohl J, Ju LK (2016) Rhamnolipid adsorption in soil: factors, unique features, and considerations for use as green antizoosporic agents. J Agric Food Chem 64 (17):3330–3337

29. Goswami D, Borah SN, Lahkar J, Handique PJ, Deka S (2015) Antifungal properties of rhamnolipid produced by Pseudomonas aeruginosa DS9 against Colletotrichum falcatum. J Basic Microbiol 55(11):1265–1274

30. De Jonghe K, De Dobbelaere I, Sarrazyn R, Höfte M (2005) Control of brown root rot caused by Phytophthora cryptogea in the hydroponic forcing of witloof chicory (Cichorium intybus var. foliosum) by means of a nonionic surfactant. Crop Prot 24(9):771–778

31. Zhao Z, Wang Q, Wang K, Brian K, Liu C, Gu Y (2010) Study of the antifungal activity of Bacillus vallismortis ZZ185 in vitro and identification of its antifungal components. Bioresour Technol 101(1):292–297

32. Haba E, Pinazo A, Jauregui O, Espuny MJ, Infante MR, Manresa A (2003) Physicochemical characterization and antimicrobial properties of rhamnolipids produced by Pseudomonas aeruginosa 47T2 NCBIM 40044. Biotechnol Bioeng 81(3):316–322

33. Kim BS, Lee JY, Hwang BK (2000a) In vivo control and in vitro antifungal activity of rhamnolipid B, a glycolipid antibiotic, against Phytophthora capsici and Colletotrichum orbiculare. Pest Manag Sci Formerly Pesticide Sci 56(12):1029–1035

34. Kim SH, Lim EJ, Lee SO, Lee JD, Lee TH (2000b) Purification and characterization of biosurfactants from Nocardia sp. L-417. Biotechnol Appl Biochem 31(3):249–253

35. Dashtbozorg SS, Miao S, Ju LK (2015) Rhamnolipids as environmentally friendly biopesticide against plant pathogen. Environ Prog 28:404–409

36. Il Kim P, Chung KC (2004) Production of an antifungal protein for control of Colletotrichum lagenarium by Bacillus amyloliquefaciens MET0908. FEMS Microbiol Lett 234(1):177–183

37. Hathout Y, Ho YP, Ryzhov V, Demirev P, Fenselau C (2000) Kurstakins: a new class of Lipopeptides isolated from Bacillus t huringiensis. J Nat Prod 63(11):1492–1496

38. Kita N, Ohya T, Uekusa H, Nomura K, Manago M, Shoda M (2005) Biological control of damping-off of tomato seedlings and cucumber Phomopsis root rot by Bacillus subtilis RB14-C. Japan Agric Res Quart JARQ 39(2):109–114

39. Mohammadipour M, Mousivand M, Salehi Jouzani G, Abbasalizadeh S (2009) Molecular and biochemical characterization of Iranian surfactin-producing Bacillus subtilis isolates and

evaluation of their biocontrol potential against Aspergillus flavus and Colletotrichum gloeosporioides. Can J Microbiol 55(4):395–404

40. Lin HF, Chen TH, Da Liu S (2011) The antifungal mechanism of Bacillus subtilis against Pestalotiopsis eugeniae and its development for commercial applications against wax apple infection. Afr J Microbiol Res 5(14):1723–1728

41. Tareq FS, Lee MA, Lee HS, Lee JS, Lee YJ, Shin HJ (2014) Gageostatins A–C, antimicrobial linear lipopeptides from a marine Bacillus subtilis. Mar Drugs 12(2):871–885

42. Andersen JB, Koch B, Nielsen TH, Sørensen D, Hansen M, Nybroe O et al (2003) Surface motility in Pseudomonas sp. DSS73 is required for efficient biological containment of the root-pathogenic microfungi Rhizoctonia solani and Pythium ultimum. Microbiology 149(1):37–46

43. Tran H, Ficke A, Asiimwe T, Höfte M, Raaijmakers JM (2007) Role of the cyclic lipopeptide massetolide a in biological control of Phytophthora infestans and in colonization of tomato plants by Pseudomonas fluorescens. New Phytol 175(4):731–742

44. Varnier AL, Sanchez L, Vatsa P, Boudesocque L, Garcia-Brugger ANGELA, Rabenoelina F et al (2009) Bacterial rhamnolipids are novel MAMPs conferring resistance to Botrytis cinerea in grapevine. Plant Cell Environ 32(2):178–193

45. Berry CL, Brassinga AKC, Donald LJ, Fernando WD, Loewen PC, de Kievit TR (2012) Chemical and biological characterization of sclerosin, an antifungal lipopeptide. Can J Microbiol 58(8):1027–1034

46. Alvarez F, Castro M, Principe A, Borioli G, Fischer S, Mori G, Jofre E (2012) The plant-associated Bacillus amyloliquefaciens strains MEP218 and ARP23 capable of producing the cyclic lipopeptides iturin or surfactin and fengycin are effective in biocontrol of sclerotinia stem rot disease. J Appl Microbiol 112(1):159–174

47. Miao S, Dashtbozorg SS, Callow NV, Ju LK (2015) Rhamnolipids as platform molecules for production of potential anti-zoospore agrochemicals. J Agric Food Chem 63(13):3367–3376

48. Lahkar J, Borah SN, Deka S, Ahmed G (2015) Biosurfactant of Pseudomonas aeruginosa JS29 against Alternaria solani: the causal organism of early blight of tomato. BioControl 60 (3):401–411

49. Singh R, Glick BR, Rathore D (2018a) Biosurfactants as a biological tool to increase micronutrient availability in soil: a review. Pedosphere 28(2):170–189

50. Sharma V, Garg M, Devismita T, Thakur P, Henkel M, Kumar G (2018) Preservation of microbial spoilage of food by biosurfactantbased coating. Asian J. Pharm. Clin. Res 11(2):98

51. Singh J, Sharma D, Kumar G, Sharma NR (eds) (2018b) Microbial bioprospecting for sustainable development. Springer, Singapore

52. Sharma A, Jansen R, Nimtz M, Johri BN, Wray V (2007) Rhamnolipids from the rhizosphere bacterium Pseudomonas sp. GRP3 that reduces damping-off disease in chilli and tomato nurseries. J Nat Prod 70(6):941–947

53. Kulakovskaya TV, Shashkov AS, Kulakovskaya EV, Golubev WI (2005) Ustilagic acid secretion by Pseudozyma fusiformata strains. FEMS Yeast Res 5(10):919–923

54. Edosa TT, Jo YH, Keshavarz M, Han YS (2018) Biosurfactants: production and potential application in insect pest management. Trends Entomol 14:79

55. Rostás M, Blassmann K (2009) Insects had it first: surfactants as a defence against predators. Proc R Soc B Biol Sci 276(1657):633–638

56. Al-qwabah AA, Al-limoun MO, Al-Mustafa AH, Al-Zereini WA (2018) Bacillus atrophaeus A7 crude Chitinase: characterization and potential role against Drosophila melanogaster larvae. Jordan J Biol Sci 11(4)

57. Yun DC, Yang SY, Kim YC, Kim IS, Kim YH (2013) Identification of surfactin as an aphicidal metabolite produced by Bacillus amyloliquefaciens G1. J Korean Soc Appl Biol Chem 56(6):751–753

58. Khedher SB, Boukedi H, Kilani-Feki O, Chaib I, Laarif A, Abdelkefi-Mesrati L, Tounsi S (2015) Bacillus amyloliquefaciens AG1 biosurfactant: putative receptor diversity and histopathological effects on Tuta absoluta midgut. J Invertebr Pathol 132:42–47

59. do Nascimento Silva J, Mascarin GM, de Paula Vieira de Castro R, Castilho LR, Freire DM (2019) Novel combination of a biosurfactant with entomopathogenic fungi enhances efficacy against Bemisia whitefly. Pest Manag Sci 75(11):2882–2891

60. Yang SY, Lim DJ, Noh MY, Kim JC, Kim YC, Kim IS (2017) Characterization of biosurfactants as insecticidal metabolites produced by Bacillus subtilis Y9. Entomol Res 47 (1):55–59

61. Moss MO (2008) Fungi, quality and safety issues in fresh fruits and vegetables. J Appl Microbiol 104(5):1239–1243

62. Irtwange SV (2006) Application of biological control agents in pre-and postharvest operations. In: Agricultural Engineering International: CIGR Journal. CIGR, Paris

63. Bi Y, Li Y, Ge Y (2007) Induced resistance in postharvest fruits and vegetables by chemicals and its mechanism. Stewart Postharvest Rev 3(6):1–7

64. Yang B, Yongcai L, Yonghong G, Yi W (2009) Induced resistance in melons by elicitors for the control of postharvest diseases. In: Postharvest pathology. Springer, Dordrecht, pp 31–41

65. Ma LY, Bi Y, Zhang ZK, Zhao L, An L, Ma KQ (2004) Control of pre-and postharvest main diseases on melon variety Yindi with preharvest azoxystrobin spraying. J Gansu Agric Univ 39:14–17

66. Marín A, Oliva J, Garcia C, Navarro S, Barba A (2003) Dissipation rates of cyprodinil and fludioxonil in lettuce and table grape in the field and under cold storage conditions. J Agric Food Chem 51(16):4708–4711

67. Rial-Otero R, Arias-Estévez M, López-Periago E, Cancho-Grande B, Simal-Gándar J (2005) Variation in concentrations of the fungicides tebuconazole and dichlofluanid following successive applications to greenhouse-grown lettuces. J Agric Food Chem 53:4471–4475

68. Leelasuphakul W, Hemmanee P, Chuenchitt S (2008) Growth inhibitory properties of Bacillus subtilis strains and their metabolites against the green mold pathogen (Penicillium digitatum Sacc.) of citrus fruit. Postharvest Biol Technol 48(1):113–121

69. Fiddaman PJ, Rossall S (1993) The production of antifungal volatiles by Bacillus subtilis. J Appl Bacteriol 74(2):119–126

70. Jiang YM, Zhu XR, Li YB (2001) Postharvest control of litchi fruit rot by Bacillus subtilis. LWT-Food Sci Technol 34(7):430–436

71. Kim PI, Bai H, Bai D, Chae H, Chung S, Kim Y et al (2004) Purification and characterization of a lipopeptide produced by Bacillus thuringiensis CMB26. J Appl Microbiol 97(5):942–949

72. Knox OGG, Killham K, Leifert C (2000) Effects of increased nitrate availability on the control of plant pathogenic fungi by the soil bacterium Bacillus subtilis. Appl Soil Ecol 15(2):227–231

73. Pinchuk IV, Bressollier P, Sorokulova IB, Verneuil B, Urdaci MC (2002) Amicoumacin antibiotic production and genetic diversity of Bacillus subtilis strains isolated from different habitats. Res Microbiol 153(5):269–276

74. Shoda M (2000) Bacterial control of plant diseases. J Biosci Bioeng 89(6):515–521

75. Wang Y, Xu Z, Zhu P, Liu Y, Zhang Z, Mastuda Y et al (2010) Postharvest biological control of melon pathogens using Bacillus subtilis EXWB1. J Plant Pathol:645–652

76. Pretorius D, Van Rooyen J, Clarke KG (2015) Enhanced production of antifungal lipopeptides by Bacillus amyloliquefaciens for biocontrol of postharvest disease. New Biotechnol 32 (2):243–252

77. Yan F, Xu S, Chen Y, Zheng X (2014) Effect of rhamnolipids on Rhodotorula glutinis biocontrol of Alternaria alternata infection in cherry tomato fruit. Postharvest Biol Technol 97:32–35

78. Imtiaz M, Rashid A, Khan P, Memon MY, Aslam M (2010) The role of micronutrients in crop production and human health. Pak J Bot 42(4):2565–2578

79. Diacono M, Montemurro F (2015) Effectiveness of organic wastes as fertilizers and amendments in salt-affected soils. Agriculture 5(2):221–230

80. Rajamani S, Gopinath M, Reddy KHP (2014) Combining ability for seed cotton yield and fibre characters in upland cotton (Gossypium hirsutum L). J Cotton Res Develop 28(2):207–210

81. Knez M, Graham RD (2013) The impact of micronutrient deficiencies inThe impact of micronutrient deficiencies in agricultural soils and crops on the nutritional health of humans. In: Essentials of Medical Geology. Springer, Dordrecht, pp 517–533
82. Miller DD, Welch RM (2013) Food system strategies for preventing micronutrient malnutrition. Food Policy 42:115–128
83. Fageria NK, Baligar VC (2003) Fertility management of tropical acid soils for sustainable crop production. In: Handbook of soil acidity, pp 359–385
84. Sinha RK, Bharambe G, Ryan D (2008) Converting wasteland into wonderland by earthworms—a low-cost nature's technology for soil remediation: a case study of vermiremediation of PAHs contaminated soil. Environmentalist 28(4):466–475
85. Mulligan CN (2005) Environmental applications for biosurfactants. Environ Pollut 133 (2):183–198
86. Mulligan CN, Yong RN, Gibbs BF (2001) Surfactant-enhanced remediation of contaminated soil: a review. Eng Geol 60(1–4):371–380
87. Ehrhardt, D. D. (2015). Produção de biossurfactante por Bacillus subtilis utilizando resíduo do processsamento do abacaxi como substrato
88. Mnif I, Ghribi D (2016) Glycolipid biosurfactants: main properties and potential applications in agriculture and food industry. J Sci Food Agric 96(13):4310–4320
89. Pacwa-Płociniczak M, Płaza GA, Piotrowska-Seget Z, Cameotra SS (2011) Environmental applications of biosurfactants: recent advances. Int J Mol Sci 12(1):633–654
90. Abdul AS, Gibson TL, Rai DN (1990) Selection of surfactants for the removal of petroleum products from shallow sandy aquifers. Groundwater 28(6):920–926
91. Bustamante M, Duran N, Diez MC (2012) Biosurfactants are useful tools for the bioremediation of contaminated soil: a review. J Soil Sci Plant Nutr 12(4):667–687
92. Olaniran AO, Balgobind A, Pillay B (2013) Bioavailability of heavy metals in soil: impact on microbial biodegradation of organic compounds and possible improvement strategies. Int J Mol Sci 14(5):10197–10228
93. Saharan BS, Sahu RK, Sharma D (2011) A review on biosurfactants: fermentation, current developments and perspectives. Genet Eng Biotechnol J 2011(1):1–14
94. Singh P, Cameotra SS (2004) Potential applications of microbial surfactants in biomedical sciences. Trends Biotechnol 22(3):142–146
95. Wang S, Mulligan CN (2004) Rhamnolipid foam enhanced remediation of cadmium and nickel contaminated soil. Water Air Soil Pollut 157(1–4):315–330
96. Gregory I (2006) The role of input vouchers in pro-poor growth. In: Background Paper Prepared for the African Fertilizer Summit, pp 9–13
97. Sheng X, He L, Wang Q, Ye H, Jiang C (2008) Effects of inoculation of biosurfactant-producing Bacillus sp. J119 on plant growth and cadmium uptake in a cadmium-amended soil. J Hazard Mater 155(1–2):17–22
98. Stacey SP, McLaughlin MJ, Çakmak I, Hettiarachchi GM, Scheckel KG, Karkkainen M (2008) Root uptake of lipophilic zinc− rhamnolipid complexes. J Agric Food Chem 56 (6):2112–2117
99. Melber C, Kielhorn J, Mangelsdorf I, World Health Organization (2004) Coal tar creosote. World Health Organization, Geneva
100. Li CH, Wong YS, Tam NFY (2010) Anaerobic biodegradation of polycyclic aromatic hydrocarbons with amendment of iron (III) in mangrove sediment slurry. Bioresour Technol 101(21):8083–8092
101. Li X, Wu Y, Lin X, Zhang J, Zeng J (2012) Dissipation of polycyclic aromatic hydrocarbons (PAHs) in soil microcosms amended with mushroom cultivation substrate. Soil Biol Biochem 47:191–197
102. Hu J, Nakamura J, Richardson SD, Aitken MD (2012) Evaluating the effects of bioremediation on genotoxicity of polycyclic aromatic hydrocarbon-contaminated soil using genetically engineered, higher eukaryotic cell lines. Environ Sci Technol 46(8):4607–4613

103. Zhu H, Aitken MD (2010) Surfactant-enhanced desorption and biodegradation of polycyclic aromatic hydrocarbons in contaminated soil. Environ Sci Technol 44(19):7260–7265

104. Szulc A, Ambrożewicz D, Sydow M, Ławniczak Ł, Piotrowska-Cyplik A, Marecik R, Chrzanowski Ł (2014) The influence of bioaugmentation and biosurfactant addition on bioremediation efficiency of diesel-oil contaminated soil: feasibility during field studies. J Environ Manag 132:121–128

105. Bezza FA, Chirwa EMN (2017) The role of lipopeptide biosurfactant on microbial remediation of aged polycyclic aromatic hydrocarbons (PAHs)-contaminated soil. Chem Eng J 309:563–576

106. Xia W, Du Z, Cui Q, Dong H, Wang F, He P, Tang Y (2014) Biosurfactant produced by novel Pseudomonas sp. WJ6 with biodegradation of n-alkanes and polycyclic aromatic hydrocarbons. J Hazard Mater 276:489–498

107. Obayori OS, Ilori MO, Adebusoye SA, Oyetibo GO, Omotayo AE, Amund OO (2009) Degradation of hydrocarbons and biosurfactant production by Pseudomonas sp. strain LP1. World J Microbiol Biotechnol 25(9):1615–1623

108. Ferradji FZ, Mnif S, Badis A, Rebbani S, Fodil D, Eddouaouda K, Sayadi S (2014) Naphthalene and crude oil degradation by biosurfactant producing Streptomyces spp. isolated from Mitidja plain soil (north of Algeria). Int Biodeterior Biodegradation 86:300–308

109. Sun S, Wang Y, Zang T, Wei J, Wu H, Wei C et al (2019) A biosurfactant-producing Pseudomonas aeruginosa S5 isolated from coking wastewater and its application for bioremediation of polycyclic aromatic hydrocarbons. Bioresour Technol 281:421–428

110. Mehetre GT, Dastager SG, Dharne MS (2019) Biodegradation of mixed polycyclic aromatic hydrocarbons by pure and mixed cultures of biosurfactant producing thermophilic and thermotolerant bacteria. Sci Total Environ 679:52–60

111. Tang Z, Zhang L, Huang Q, Yang Y, Nie Z, Cheng J et al (2015) Contamination and risk of heavy metals in soils and sediments from a typical plastic waste recycling area in North China. Ecotoxicol Environ Saf 122:343–351

112. Albuquerque CF, Luna-Finkler CL, Rufino RD, Luna JM, de Menezes CT, Santos VA, Sarubbo LA (2012) Evaluation of biosurfactants for removal of heavy metal ions from aqueous effluent using flotation techniques. Int Rev Chem Eng 4(2):156–161

113. Luna JM, Rufino RD, Sarubbo LA (2016) Biosurfactant from Candida sphaerica UCP0995 exhibiting heavy metal remediation properties. Process Saf Environ Prot 102:558–566

114. Vijayakumar S, Saravanan V (2015) Biosurfactants-types, sources and applications. Res J Microbiol 10(5):181

115. Akbari S, Abdurahman NH, Yunus RM, Fayaz F, Alara OR (2018) Biosurfactants—a new frontier for social and environmental safety: a mini review. Biotechnol Res Innov 2(1):81–90

116. Guo X, Wei Z, Wu Q, Li C, Qian T, Zheng W (2016) Effect of soil washing with only chelators or combining with ferric chloride on soil heavy metal removal and phytoavailability: field experiments. Chemosphere 147:412–419

117. Park J, Son Y, Noh S, Bong T (2016) The suitability evaluation of dredged soil from reservoirs as embankment material. J Environ Manag 183:443–452

118. Haryanto B, Chang CH (2015) Removing adsorbed heavy metal ions from sand surfaces via applying interfacial properties of rhamnolipid. J Oleo Sci 64(2):161–168

119. Santos DK, Resende AH, de Almeida DG, Soares da Silva RDCF, Rufino RD, Luna JM et al (2017) Candida lipolytica UCP0988 biosurfactant: potential as a bioremediation agent and in formulating a commercial related product. Front Microbiol 8:767

120. Diaz MA, De Ranson IU, Dorta B, Banat IM, Blazquez ML, Gonzalez F et al (2015) Metal removal from contaminated soils through bioleaching with oxidizing bacteria and rhamnolipid biosurfactants. Soil Sediment Contam Int J 24(1):16–29

121. Yang Z, Zhang Z, Chai L, Wang Y, Liu Y, Xiao R (2016) Bioleaching remediation of heavy metal-contaminated soils using Burkholderia sp. Z-90. J Hazard Mater 301:145–152

122. Aşçı Y, Nurbaş M, Açıkel YS (2008) Removal of zinc ions from a soil component Na-feldspar by a rhamnolipid biosurfactant. Desalination 223(1–3):361–365

123. Alternative Remedial Technologies Inc. (1992) Engineering. "Soil Washing at King of Prussia Superfund Site"
124. Ferraro A, van Hullebusch ED, Huguenot D, Fabbricino M, Esposito G (2015) Application of an electrochemical treatment for EDDS soil washing solution regeneration and reuse in a multi-step soil washing process: case of a cu contaminated soil. J Environ Manag 163:62–69
125. Khalid S, Shahid M, Niazi NK, Murtaza B, Bibi I, Dumat C (2017) A comparison of technologies for remediation of heavy metal contaminated soils. J Geochem Explor 182:247–268
126. Bilgin M, Tulun S (2016) Removal of heavy metals (cu, cd and Zn) from contaminated soils using EDTA and FeCl3. Global NEST J 18:98–107
127. Mulligan CN, Wang S (2006) Remediation of a heavy metal-contaminated soil by a rhamnolipid foam. Eng Geol 85(1–2):75–81
128. Singh AK, Cameotra SS (2013) Efficiency of lipopeptide biosurfactants in removal of petroleum hydrocarbons and heavy metals from contaminated soil. Environ Sci Pollut Res 20(10):7367–7376
129. Juwarkar AA, Nair A, Dubey KV, Singh SK, Devotta S (2007) Biosurfactant technology for remediation of cadmium and lead contaminated soils. Chemosphere 68(10):1996–2002
130. Banat IM, Franzetti A, Gandolfi I, Bestetti G, Martinotti MG, Fracchia L et al (2010) Microbial biosurfactants production, applications and future potential. Appl Microbiol Biotechnol 87(2):427–444
131. Farn RJ (ed) (2008) Chemistry and technology of surfactants. John Wiley & Sons, New York
132. Gogoi SB (2011) Adsorption–desorption of surfactant for enhanced oil recovery. Transp Porous Media 90(2):589
133. Al-Sulaimani H, Joshi S, Al-Wahaibi Y, Al-Bahry S, Elshafie A, Al-Bemani A (2011) Microbial biotechnology for enhancing oil recovery: current developments and future prospects. Biotechnol Bioinformatics Bioeng 1(2):147–158
134. Rahman PK, Gakpe E (2008) Production, characterisation and applications of biosurfactants-review. Biotechnology
135. Youssef N, Simpson DR, McInerney MJ, Duncan KE (2013) In-situ lipopeptide biosurfactant production by Bacillus strains correlates with improved oil recovery in two oil wells approaching their economic limit of production. Int Biodeterior Biodegradation 81:127–132
136. Joshi S, Bharucha C, Desai AJ (2008) Production of biosurfactant and antifungal compound by fermented food isolate Bacillus subtilis 20B. Bioresour Technol 99(11):4603–4608
137. Geetha SJ, Banat IM, Joshi SJ (2018) Biosurfactants: production and potential applications in microbial enhanced oil recovery (MEOR). Biocatal Agric Biotechnol 14:23–32
138. De Almeida JM, Miranda CR (2016) Improved oil recovery in nanopores: NanoIOR. Sci Rep 6:28128
139. El-Sheshtawy HS, Aiad I, Osman ME, Abo-ELnasr AA, Kobisy AS (2015) Production of biosurfactant from Bacillus licheniformis for microbial enhanced oil recovery and inhibition the growth of sulfate reducing bacteria. Egypt J Pet 24(2):155–162
140. Zou C, Wang M, Xing Y, Lan G, Ge T, Yan X, Gu T (2014) Characterization and optimization of biosurfactants produced by Acinetobacter baylyi ZJ2 isolated from crude oil-contaminated soil sample toward microbial enhanced oil recovery applications. Biochem Eng J 90:49–58
141. Gandler G, Gbosi A, Bryant SL, Britton LN (2006, January) Mechanistic understanding of microbial plugging for improved sweep efficiency. In: SPE/DOE Symposium on Improved Oil Recovery. Society of Petroleum Engineers, Richardson
142. Martins PC, Martins VG (2018) Biosurfactant production from industrial wastes with potential remove of insoluble paint. Int Biodeterior Biodegradation 127:10–16

Toxicity and Biodegradability Assessment

6

Abstract

Microbial surfactants are environmentally safe and can be easily degradable as the composition is mainly composed of polysaccharides, fatty acids, and protein, etc. But simultaneously surfactants have also been reported for their impact on environmental disposal, irritants when coming to the cosmetic formulations, and safety in terms of food safety. The toxicity of the concerned surfactants depends upon the hydrophobicity of the molecule, the higher toxicity has been displayed by highly hydrophobic surfactants. On the contrary, the toxicity of microbial surfactants fate has been assessed in the environment and industrial formulation has been studied and modeled up to an extent. That is why various industries are adopting and looking for microbial surfactants as a substitute for synthetic surfactants. So, it is essential to increase the investigation about such concerns to elucidate such aspects, which are essential to get accurate safety data about their adoption to the industry. The major concerns and regulations about the surfactants are biodegradability, toxicity, and impact on human health. The adoption of microbial surfactants could expand all such aspects in comparison with synthetic surfactants and should be well-thought-out in the predictable future.

Keywords

Cytotoxicity · Phytotoxicity · Cell lines · Non-toxic · Irritants

6.1 Introduction

BSs are composed of diverse chemical compositions, comparatively low toxicity when compared to synthetic surfactants, and their stability in extreme environmental conditions such as high temperature, functionality in extended pH ranges, ionic

© Springer Nature Singapore Pte Ltd. 2021
D. Sharma, *Biosurfactants: Greener Surface Active Agents for Sustainable Future*,
https://doi.org/10.1007/978-981-16-2705-7_6

strength, and metal concentrations [1–5]. Microbial surfactants are environmentally safe and can be easily degradable as the composition is mainly composed of polysaccharides, fatty acids, protein, etc. The domain of surfactants research and utilization has manifested its presence in almost all types of applications and aspects of human life such as household chemicals and cleaning solutions, cosmetics formulations, pesticide formulations, pharmaceuticals, and other petrochemicals applications [6]. The range of biosurfactants-based formulation extended from mild personal care products such as toothpaste, cosmetic gels, and soaps to high concentration applications like environmental, agricultural, and as an emulsifier in microbial enhanced oil recovery (MEOR). The surfactant market is anticipated to grow at a targeted value of \$44.9 billion by the end of 2022.

Besides, BSs may be explored in various plant growth promotion, improvement in the soil quality by enhancing micronutrient availability, and environmental remediations. It has been demonstrated that the BSs have less toxic to the ecological components than chemical surfactants [7, 8, 9]. But simultaneously surfactants have also been reported for their impact on environmental disposal, irritants when coming to the cosmetic formulations, and safety in terms of food safety. Surfactants from different reports and studies anticipated that surfactants have seriously depleted and impaired the macro and microbiota of the soil and aquatic environment [10]. The surfactant toxicity is due to its improper disposal to nearby water bodies and terrestrial sites [11]. Not only adjoining water bodies and sites, but humans health is also seriously pretentious by the consequences of surfactant pollution [12]. Mostly, all the surfactants obtained and produced can be categorized based on the origin of the molecule, viz., petrochemical surfactants, oleochemicals derived from chemicals, or biosurfactants obtained from microbial and plant cells as portrayed in Fig. 6.1.

The utilization of petrochemical and oleochemical synthesized surfactants has attained superiority over biosurfactants because of their production economics. Although the financial expenses incurred due to the petrochemical and oleochemical process is also noteworthy.

The sodium lauryl a type anionic surfactant is one of the key constituents of toothpaste. Also, being a foaming compound, sodium lauryl converses with sensory characteristics and has an added non-required antimicrobial potential. Though, the common utilization of such chemical-based surfactants can cause different types of irritations, allergic like dermatitis, inflammation, irritation of the mucous surfaces, and ulcerations [13, 14]. Furthermore, the consumption of sodium lauryl sulfate can result in carcinogenic effects [15].

Most of the chemical surfactants accumulated mostly in the wastewater, which further ends up in the local waterbodies as its discharged and pretense a hazard to the local ecosystems [16, 17]. The extent of the hazard depends on the type and concentration accumulated for the particular surfactants to the water bodies and environmental sites and also depends upon the exposure time. Most of the microorganisms associated with the natural cleaning and breakdown of the

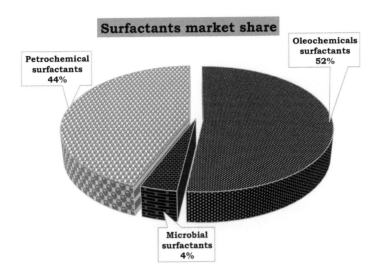

Fig. 6.1   An outline of surfactants market share and its origin

contaminants also get affected because of the cell membrane disintegration due to the presence of a high concentration of accumulated surfactants [18].

The toxicity of the concerned surfactants depends upon the hydrophobicity of the molecule, the higher toxicity has been displayed by highly hydrophobic surfactants. The incidence of chemical surfactants in the aquatic environment in the last decade has to lead to excessive toxicity. Furthermore, such accumulation of the surfactants results in the biomagnification of such molecules to the higher trophic level such as aquatic animals and weeds used for food and feed purposes. The biomagnification of surfactants can impair enzyme activity. The inadequate information about the toxicity assessment is the major hindrances to the growth of the surfactant market [19]. On the contrary, the toxicity of microbial surfactants fate has been assessed in the environment and industrial formulation has been studied and modeled up to an extent. That is why various industries are adopting and looking for microbial surfactants as a substitute for synthetic surfactants.

On the contrary, BSs have been rising interest in the utilization of greener substitutes and highly biocompatible alternatives for petrochemical and oleochemicals origin surfactants [20]. Biodegradability of the microbial surfactants was detected as an extent (%)and determined based on the increase in surface tension of samples, with a range of 80% or more biodegradability in specific conditions. Moldes et al. [21] demonstrated in a study that bioremediation of the octane in soil system by exploring the BSs derived from the *Lactobacillus pentosus*. BSs derived from the LAB reported for about 60% after 15 days of treatment, and 76% octane degradation after incubation of 30 days. Whereas the rate of degradation of octane was found approximately three times lower. The role of the BSs in the degradation of the hexachlorocyclohexane (HCH) was reported as an efficient approach [22]. BSs were utilized as an active fraction of the bioremediation media for HCH degradation. Developing environmentally friendly, non-allergic, non-irritant, mild on non-toxic, and potentially effective surfactants for the food and healthcare sector is a continuing challenge in the industry. The eco-friendly and inexpensive surface-active agents are still a challenging area in industrial formulations.

6.2   Biodegradability of Surfactants

6.2.1   Comparative Life Cycle Assessment (LCA) of Surfactants

Life cycle assessment of any substance is an approach for evaluating environmental influences related to all the phases of the life cycle of an industrial product, progression, or service. LCA of any procedure plays a noteworthy role in evaluating the complete process of any natural or manmade material from starting to raw materials, different phases of the processing, and finally the consequences of the synthesis on the ecology and environment. LCA, every so often called the "cradle to grave" and from time to time circumscribed to "cradle to gate" investigative strategy of production schemes, expands the restrictions of ecological impact assessment [23].

The industrial production of the chemical-based surfactants exclusively depends on the modification of various petroleum substances obtained from by-products to surface-active agents. The surfactants can be categorized as anionic, cationic, non-ionic, and amphoteric in nature based on the net charge of the molecule. The largely utilized laundry surfactants are commonly synthesized as a blend of an extensive display of surfactant agents along with enzyme powder to raise their cleaning performance (Fig. 6.2). In a report, Thannimalay and Yusoff [24] described that the LCA of linear alkylbenzene sulfonate, which is one of the largely utilized chemical anionic surfactants, has been observed to considerably disturb the production of fossil fuels, change the degree of land usage, and leads to the accumulation of respiratory inorganics [24].

Microbial surfactants are considered green surface-active agents due to their environmental friendliness, less toxicity, and biodegradability, as compared to chemical surfactants. LCA assessment of the microbial surfactants has been depicted in Fig. 6.3. A comparative investigation of various BSs, like glycolipids, specifies that the key contributors to environmental effects include air emissions, electricity consumption, and thermal necessities [25]. It was evident by Baccile et al. [26] that the synthesis of acetylated sophorolipids is observed to be effective than the sophorolipids in comparison to the toxicity inhibited by them (Figs. 6.4 and 6.5).

The cradle-to-grave assessment was achieved for a hand-washing application, which permitted the fortitude of hot spots and contrast to reference substances on the market. This LCA strategy exploring the same substance batch is extremely innovative in the domain of BSs, where inconsistency among outcomes in the literature is every so often attributable to expected variations in the BSs constituents utilized for

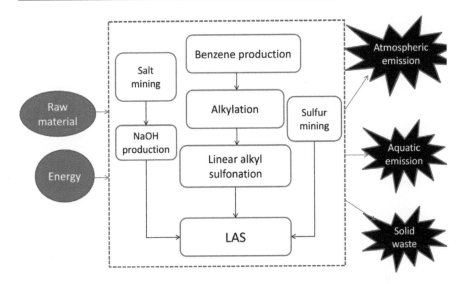

Fig. 6.2 LCA outlines for the production of synthetic surfactants, i.e. Linear alkylbenzene sulfonate (Courtesy by: [9])

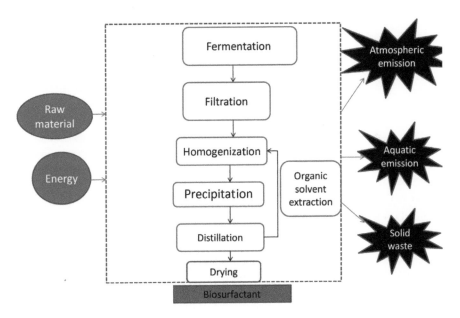

Fig. 6.3 LCA outlines for the production of microbial surfactants (Courtesy by: [9])

the experiments, often combined with the poor portrayal, making it problematic to draw consistent conclusions.

LCA was observed to regulate the effect of such a new production bioprocess. The environmental effect was unexpectedly comparable to that of synthetic

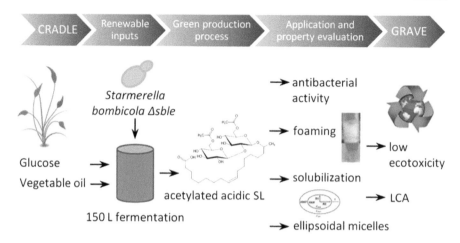

Fig. 6.4 LCA or Cradle-to-Grave strategy for acetylated acidic sophorolipid surfactants (Courtesy by: [26])

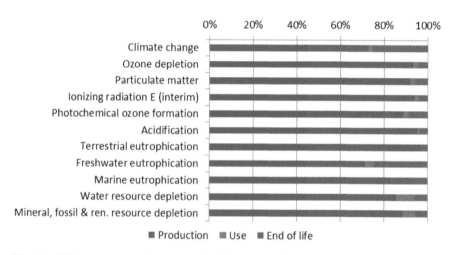

Fig. 6.5 LCA hot spots for acidic sophorolipids in hand-washing application including the whole bioprocess (Courtesy by: [26])

surfactants obtained using fossil resources. It was observed that the utilization of glucose and rapeseed oil as the key nutrient is the major cause of approximately 78% damage score to the environment. The residual influence was mainly due to electricity usage during the process, i.e. 15%. The optimization of the bioprocess and substrate conversion efficiency, the compositional optimization, and modeling for the utilization of glucose: and rapeseed oil holds the key improvement aspect. The outcome of the above-discussed study opens the result of the way expansion of a cradle-to-grave approach that is green, accessible, and applicable for different industries for the production of well-identified BSs with extraordinary potential.

Fig. 6.6 Life cycle assessment process (ISO Standard 14,044)

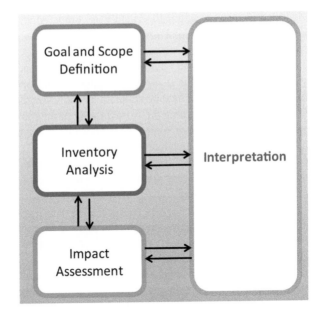

The outline of LCA was planned by the International Standards Organization (ISO) relating to the ISO Standard 14,040 with four different operational steps, viz., (1) goal and scope, (2) inventory assessment, (3) impact analysis, and (4) interpretation (ISO Standard 14,040; Brusseau, 2019) as detailed in Fig. 6.6. An LCA assessment of the production of palm oil derived biobased surfactant, i.e. methyl ester sulfonates (MES) specified that key impact groups fossil fuel, respiratory organic compounds, and climate change and were observed to be relatively less when linked to the influence instigated by petroleum-based surfactants (Zolkarnain et al., 2016). In the same report, it was observed that the biobased surfactants are more eco-friendly as compared to the LAS after LCA assessment throughout the production pathway [24]. The result of such studies opens the aftermath of the way the growth of a cradle-to-grave strategy that is green, available, and appropriate for various industries for the synthesis of well-recognized biobased surfactants with astonishing potential.

Kopsahelis et al. [25] demonstrated gate-to-gate LCA of BSs and bio plasticizers synthesis using different fats and similar waste. The study determined that the consumption of electricity, the requirement of thermal, the quantity of air emissions are the major contributors that impact the LCA pathways of any products. It was observed that sophorolipids are required more energy as compared to the production of rhamnolipids which ultimately results in lower ecological impacts. The amplified thermal necessities of the monoglycerides production stage are the key responsibilities to impact on the environment in negative directions and performance, which is 3 times higher when compared to the monoglycerides with fatty acid ethyl esters. In a conclusion, it was determined that in all the BS, sophorolipids synthesis caused 22.7% higher environmental influence when compared with other glycolipids such as rhamnolipids. Likewise, fatty acid ethyl esters synthesis was identified as more ecologically friendly as compared with monoglycerides, ensuing in 67% less environmental impact grounded on electricity consumption, thermal energy, and air pollution.

On the other front, dynamic LCA assessment of glycolipids such as sophorolipids production using waste valorization opens one more paradigm that can be a workout and should be adopted. There is very limited information available about the studies concerning the sustainability of microbial surfactants. The prospect related to the LCA assessment of the sophorolipids has been carried out by Hu et al. [27]. In that attempt, a dynamic LCA assessment of the sophorolipids has been proposed and carried out the framework that quantifies the environmental impacts iteratively, in sophorolipids synthesis. The dynamic LCA assessment involved the selection of suitable feedstock, production strategies, and in-situ integrated downstream processing and recovery process, give rise to lower environmental impacts. Such kind of improvement would surely result in the environmental sustainability of BSs synthesis. Subsequent data analysis and factsheets can be iterative to the purpose of technological variations and alleviate the associated impacts.

The LCA assessment of the biotechnological products and metabolites provides close insight and into its impact on the environment. Though, the confidence trustworthiness of such assessment critically depends on the effective enclosure of human toxicity and environmental impacts. The selection of appropriate proxy variables and effective LCA analytical proficiency becomes very vital. Thus the selection of improved and environmentally friendly solutions to surface-active agents synthesis should be implemented for suitable environmental sustainability.

6.2.2 Toxicity and Degradative Comparison of Surfactants and Biosurfactants

The majority of the microbial surface-active agents are obtained from bacteria, actinomycetes, yeasts, and molds during cultivation on different carbon sources [28]. BSs have indistinct advantages over synthetic surfactants, which include lower toxicity, a high degree of biodegradability, and improved ecological compatibility. Biodegradability is recognized as a vital feature, exclusion of these amphiphiles from environmental sites, and, henceforth, degradability is regarded as an important property when estimating the environmental hazard linked with surfactant use [29]. Biosurfactants are generally composed of sugar, lipid, amino acids, and proteins with phosphates. All these biological macromolecules are degradable by indigenous microflora of any ecological niche.

The environmental breakdown of the BSs molecules involves two steps when microorganisms are involved and utilize it as carbon and energy sources. Initially, hydrocarbon breakdown occurs, leading to structural alteration and instantaneous forfeiture of amphiphilicity. Consequently, the intermediate product resulting from the first step is further transformed into CO_2, water, and elemental minerals (Eq. 1 and 2) [30].

$$BSs + \text{degradative microorganisms} = Sugar + Lipids + \text{Amino acids} \quad (6.1)$$

$$Sugar + Lipids + \text{Amino acids} = CO_2 + H_2O + minerals \quad (6.2)$$

Degradation assessment is also quite critical along with the toxicity evaluation, especially BSs in aquatic and soil systems. So, altogether, the degree of biodegradation and toxicity are the information needed to observe the possible impact of such substances on the biota of a selected ecosystem and to adjust the utilization of such substances [31]. The key well-known group of internationally recognized biodegradation experimentation is the Guidelines for the Testing of Chemicals and impact on environment planned by the Organization for Economic Co-operation and Development (OECD) [32].

The test was conducted under the aegis of Organization for Economic Cooperation and Development (OECD) 301D for complete biodegradability. Degradation of a surfactant and the extent of degradation impacts either positive and negative outcomes on its utilization in bioremediation activities [33]. The depletion of oxygen, minerals, and intermediates by-product accumulated during degradation are some of the negative impacts [34, 35]. The removal or degradation of surfactants from terrestrial and aquatic systems is one of the positive effects of bioremediation. At the same time, various surfactants such as linear alkylbenzene sulfonate are utilized in remediation and can be degraded under anaerobic conditions [36]. In a report, Mohan et al. (2006) demonstrated that rhamnolipid biosurfactants can be degraded under aerobic conditions, whereas a chemical surfactant such as Triton X-100 was only partly biodegradable further down in aerobic conditions.

In a major study about the biodegradability of bacterial surfactants in a liquid medium using soil microcosms demonstrated their potential in bioremediation [37]. Various microbial surfactants have been selected for the biodegradation assay such as BSs derived from *Bacillus subtilis*, viz. surfactin, fengycin, and iturin and fengycin glycolipid obtained from *Dietzia maris*, arthrofactin obtained from *Arthrobacter oxydans*, and flavolipids obtained from Flavobacterium sp. utilized as major substrates by the *Acinetobacter haemolyticus*, *Acinetobacter baumanni*, and Pseudomonas sp. in a minimal medium. In their observation, it was evident that Pseudomonas sp. were displayed two order degree of degradation as compared to the *A. haemolyticus* and *A. baumannii* in the liquid minimal medium. Whereas *A. haemolyticus* and *A. baumannii* have been reported for the biodegradation of chemical surfactant SDS which is quite comparable to the biosurfactants used in the present study, but actually, it was an order of degree lower than BS degradation observed with Pseudomonas sp.

In soil microcosms, it was established that the microbial surfactants have been utilized as a substrate during the microbial growth of *A. baumannii*, *A. junni*, and Pseudomonas sp. The BSs did not vary in their capability to excite the respiratory activity of the consortium, but the degradation of the SDS has not been observed with mixed culture even after 7 days of incubation. It was established that the biological surfactants had a better degradation possibility as compared to the chemical surfactant SDS. BSs displayed stability in the soil system and found it companionable for ecological applications, at the same time, the significant rate of degradation reduces the chance of their buildup in the environment.

Sophorolipids obtained from the *C. bombicola* have been evaluated for cytotoxicity, degradability as compared to the chemical surfactants [38]. To explore the possibility of sophorolipids for industrial applications, the BS was evaluated based on interfacial potential, cytotoxicity, and degradability and compared such properties with lipopeptides such as surfactin and arthrofactin, sodium laurate, and various synthetic surfactants together with two block-copolymer non-ionic surfactants (BPs), polyoxyethylene lauryl ether (AE), and sodium dodecyl sulfate (SDS).

Cytotoxicity of sophorolipids on human keratinocytes was found similar to surfactin, which has previously been industrialized as a cosmetic ingredient. Furthermore, the biodegradability of sophorolipids based on the OECD testing Guidelines of Chemicals exhibited that SLs can be categorized as a "readily" biodegradable agent, which means a compound that can be degraded to the 60% extent within 28 days under specified test protocols. It was displayed that 61% degradation of sophorolipids on the eighth day of incubation. It has been concluded that sophorolipids are low-foaming surface-active agents with significant detergency, non-cytotoxicity, and efficiently biodegradable behavior.

Biosurfactants are commonly said to be biodegradable. Indeed, examining the biodegradation of SLs according to the OECD 301C method revealed that SLs are readily biodegradable chemicals. Renkin, using OECD 301F as an alternative biodegradable test, has reported that the biodegradation of SLs was sufficient to be classified as readily biodegradable chemicals [39]. This report also supports that SLs are highly biodegradable surfactants.

The extent of degradability and toxicity even depends upon the type of stereoisomers and method of synthesis (natural/tailor-made enzymatic synthesis). Mono-rhamnolipids diastereomers were reported for different degrees of degradability and toxicity [40]. The biodegradability of the chemically synthetic mono-rhamnolipids differs from microbially synthesized rhamnolipids because they are obtained as single congeners. Each congener is obtained as one of four probable diastereomers consequential from two chiral centers at the carbinols of the fatty acid tails as (R,R), (R,S), (S,R), and (S,S) (Fig. 6.7). The degradability of all the diastereomers has been compared based on CO_2 respirometry, toxicity in terms of acute toxicity, embryo-toxicity, and cytotoxicity of chemically synthesized (Rha-C10-C10) mono-rhamnolipids as compared to the microbially derived mono-rhamnolipid mixtures (bio-mRL). On the ground of biodegradability, rhamnolipid (Rha-C10-C10) diastereomers and biologically synthesized mono-rhamnolipids were found characteristically biodegradable extending from 34 to 92% mineralized. The acute toxicity of all chemically synthesized Rha-C10-C10 diastereomers and microbially derived mono-rhamnolipids are to some extent toxic as per the guidelines of US EPA ecotoxicity classification with 5 min at effective concentration ($EC_{50)}$ values extending from 39.6 to 87.5µM. The cell cytotoxicity assessment established that the inhibitory concentration (IC_{50}) values extending from 103.4 to 191.1µM for the cell lines of human lung, i.e. H1299 after 72 h of incubation.

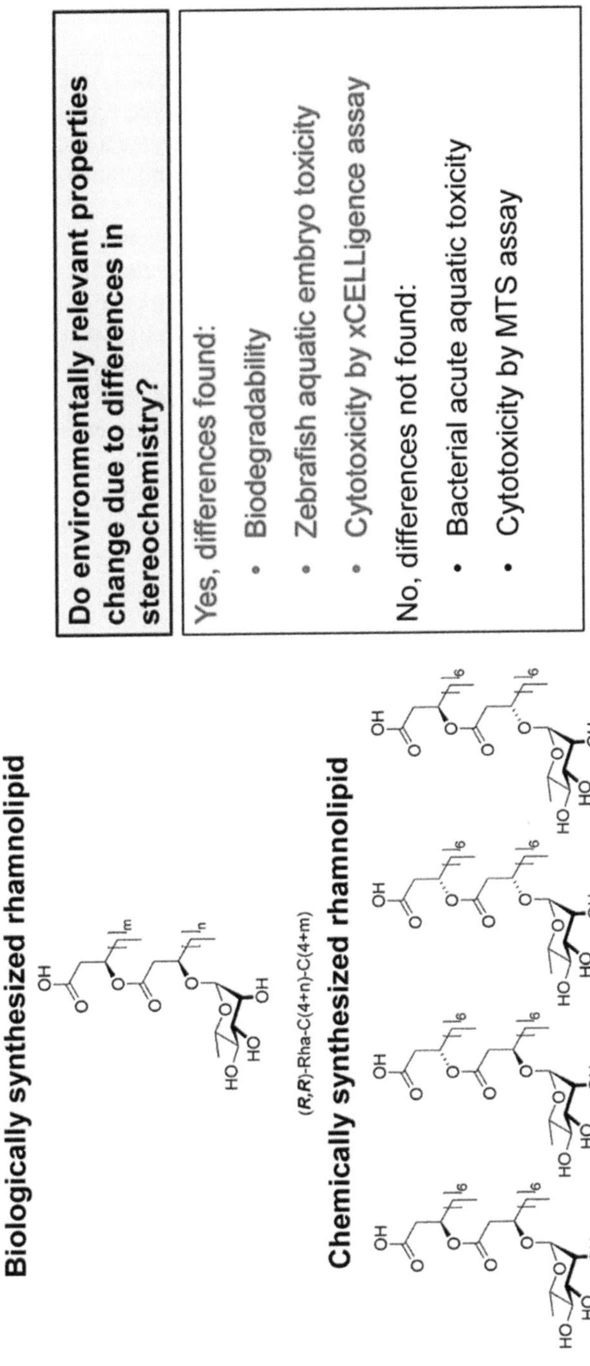

Fig. 6.7 Biodegradability and toxicity assessment of mono-rhamnolipid diastereomers (Courtesy by: [40])

Novel mono-rhamnolipids with modified stereochemistry or congener composition may have diverse properties. In such a study, that such variances resulted in measurable deviations in degradation, acute toxicity, and cell lines of the human lung; no noteworthy variances were observed for acute toxicity and cytotoxicity assay. Such comparison places a foundation for considering the ecological compatibility of chemically synthesized mono-rhamnolipid diastereomers and delivers pertinent evidence when considering possible utilization as an individual component or as a mixture.

BSs obtained from the *Bacillus subtilis* isolated from the mangrove soil in Brazil [41]. The environmental compatibility of BS derived from *Bacillus subtilis* strain was established. The acute toxicity test of the BS was carried out with *Daphnia magna*, *Vibrio fischeri*, and *Selenastrum capricornutum* specified that the BS toxicity is comparatively lower than that of its synthetically synthesized counterparts. The observation of the biodegradability evaluation confirmed that the BS can be biodegraded by *Pseudomonas putida* and other microbial cultures from a sewage treatment site. The degradation potential is reliant on the BS initial concentration.

In this work, the environmental properties of crude surfactin produced by *B. subtilis* ICA56 are reported. Acute toxicity tests carried out with *V. fischeri*, *D. magna*, and *S. capricornutum* show that the toxicity of the biosurfactant is lower than that reported for conventional surfactants. The results of the tests with *P. putida* indicate that the biodegradation process depends on the initial concentration of surfactin, the maximal degradability at 100 mg/L. In conclusion, as the degradation extent attained in the present work complies with OECD Guidelines for Testing of Chemicals (OECD 301E), surfactin can be categorized as a "readily" biodegradable substance.

Various methods and approaches have been used for the toxicity assessment of BSs as summarized in Table 6.1.

6.2.3 Toxicity Experimental Models

There are various models have been optimized and validated by the various researcher while working on toxicity assessment of microbial surfactants (as summarized in Table 6.1). No such specific guidelines are available from the regulatory agencies that have been developed and provided. But at the same time, various model studies to test food additives, cosmetics ingredients, environmental fate, and biodegradation data can be used as related information, such as Organization for Economic Co-operation and Development (OECD) guidelines for testing of chemicals (301E and GB/T 21604–2008 (Standards I, 2008a), GB/T 21603–2008 (Standards I, 2008b), and GB/T 27861–2011 (Standards I 2011).

Table 6.1 Different toxicity models assessment for biosurfactants reported

Biosurfactants	Phytotoxicity	Animal model	Ecotoxicity	Cell lines	Reference
Glycolipid	Vegetable seeds	NP	*Anomalocardia brasiliana, Artemia salina* (larvae)	NP	Santos et al. [42, 43]
Glycolipid	*Brassica oleracea, Cichorium intybus, Solanum gilo*	NP	NP	NP	de Souza Sobrinho et al. (2013)
Mono-rhamnolipid	NP	Zebrafish	Microtox assay	(H1299) human lung cell line	Hogan et al. [40]
Lipopeptide	NP	Acute toxicity on mice	NP	NP	Anyanwu et al. [44]
Lipopeptides	NP	Acute toxicity on mice (LD$_{50}$)	NP	NP	Sahnoun et al. [20]
Lipopeptides	NP	NP	*Artemia franciscana* larvae	NP	Kiran et al. [45]
Lipopeptides & Rhamnolipids	NP	NP	*Anopheles stephensi* larvae	NP	Parthipan et al. [46]
Mono and di rhamnolipid congeners	NP			Mouse L292 fibroblastic cell line	Patowary et al. [47]
Surfactin & Fengycin	NP	*Daphnia similis*		NP	Trejo-Castillo et al. [48]
Lipopeptide	NP	Acute toxicity tests, carried out with *Daphnia magna*	NP	NP	Catter et al. [49]
Surfactin	NP	Acute dermal irritation, acute oral toxicity	NP	NP	Fei et al. [50]
Glycolipid	NP	NP	Short-term acute toxicity test with *Artemia nauplii*	NP	Zenati et al. [51]
Glycolipid	*Brassica nigra* and *Triticum aestivum*	NP	NP	Mouse fibroblast (ATCC L929)	Sharma et al. [52]
Xylolipid	NP	NP	Oral and dermal toxicity in mice	NP	Saravanakumari and Mani [53]

6.2.4 Skin Irritation Test (SIT)

SIT is one of the well-known practices to detect the toxicity of any given materials to be used in healthcare, personal care, and cosmetics. In the monitoring of hazard classification and category, the framework is well-defined as the development of reversible impairment to the skin after chemical exposure. SIT is an in vitro, test with non-animal involvement defined to recognize such chemicals and substances capable of persuading judicious skin irritation (UN GHS Category 2 Skin Irritants), and to discriminate UN GHS Category 2 Skin Irritants from UN GHS 3 Mild skin irritants as well as those not needing categorization for skin irritation potential.

Fei et al. [50] performed the surfactin lipopeptides skin irritation using (LD_{50} and LC_{50}) measured as per the GB/T 21604–2008 (Standards I, 2008a), GB/T 21603–2008 (Standards I, 2008b), and GB/T 27861–2011 (Standards I, 2011). The measurement of the skin dermal irritation protocol was articulated as the primary irritation index (PII) (Table 6.2).

While working with *L. pentosus* derived BS, it has been demonstrated that did exhibit any irritant effect during the irritant testing with the chorioallantoic membrane of hen's egg assay as compared to the chemical surfactants SDS [54].

6.2.5 Acute Toxicity

Acute systemic toxicity estimates the antagonistic effects that come out with the subsequent contact of test organisms to solo or multiple doses by a known route of test material in the period of 24 h. The determination of acute toxicity is the necessary framework by the regulatory bodies for labeling and categorization of material for human utilization [55, 56]. Data obtained after acute toxicity assessment assist as a roadmap in dosage range for long-term toxicity evaluation which involves the practice of animal models [57].

The concentration of the material required to kill 50% of the test animals as described in the GB/T 21603–2008 was evaluated. LD_{50} evaluation has been demonstrated on animals by the oral route, and the outcome was expressed as the concentration of material administered per kilogram of the bodyweight of test animals (Table 6.3). The mortality rate of mice was observed at 96 h, and the amount of test material on the death of 50% of mice was observed, which is defined by LC_{50}.

Fei et al. [50] elaborated acute oral toxicity of BSs obtained from the *Bacillus subtilis* as >5000 mg kg^{-1}, LC_{50} > 1000 mg kg^{-1}) which indicates that the BS is a low-toxic ingredient.

Table 6.2 Primary irritation/dermal index (PDII)

Categorization	Primary irritation/dermal index (PII)
No irritation	0
Slight irritation	>0–2.0
Moderate irritation	2.1–5.0
Severe irritation	>5.0

Table 6.3 Categorization of LD_{50} based on dose range [58]

Categorization	Lethal dose (LD_{50})
Extremely toxic	<5 mg/kg
Highly toxic	5–50 mg/kg
Moderately toxic	50–500 mg/kg
Slightly toxic	500–5000 mg/kg
Practically non-toxic	5000–15,000 mg/kg
Relatively harmless	>15,000 mg/kg

A lipopeptide biosurfactant tested in male mice had LD_{50} of 475 mg/kg. However, daily doses of up to 47.5 mg/kg showed no negative hematological or serum biological data effects [20].

6.3   Ecotoxicity Assessment

6.3.1   Artemia Assay

The toxicity assessment of the BSs was established with the various models utilizing brine shrimp as the ecotoxicity indicator. Larvae of brine shrimp can be used for ecotoxicity assessment. The brine shrimp larvae can be tested for different concentrations of microbial BSs under static culture conditions for a short time testing (Table 6.4). The toxicity threshold concentration, defined as BS amount concentration per 100 ml of solution, is considered as the lowest amount that killed all the larval stage of brine shrimp within 24 h.

6.3.2   Anomalocardia Brasiliana

Ecotoxicity can be also performed using the *Anomalocardia brasiliana* as per the protocol described by Standard Methods for the Examination of Water and Wastewater [62]. Santos et al. [42, 43] demonstrated the toxicity of BS obtained from *Candida lipolytica* to be used as a bioremediation agent. For the assessment of acute toxicity, a test solution in the quantity of 1.0–2.0 L was sustained with constant aeration at room temperature for 12 hr. The observations of the test determined the LC_{50}, which is expressed in terms of mortality of three independent readings for each tested dilutions with the BS.

6.4   In Vitro Cytotoxic Effect

In vitro cytotoxicity assessment is recognized as screening methods, utilized to determine the living cell's behavior to the tested materials in a cell culture method, including cell viability and the ability to perform cellular growth. In vitro cytotoxicity methods utilize endpoint criteria like viability, proliferation, damage to the cell

Table 6.4 Various studies depicting the type of BSs tested for ecotoxicity against Shrimps

S. No.	Biosurfactants	Environmental component	References
1.	Glycolipids produced by *Candida lipolytica*	Micro-crustacean *Artemia salina*	Diniz Rufino et al. [59]
2.	Glycolipids obtained from *Candida sphaerica* UCP0995	*Artemia salina*	Bezerra de Souza Sobrinho et al. [60]
3.	Glycolipids of *Candida lipolytica*	*Artemia salina*	Santos et al. [42, 43]
4.	Rhamnolipids	*Artemia salina*	Deivakumari et al. [61]

Table 6.5 Equivalence between Cell Viability and Cytotoxicity [64, 65]

Cell viability (%)	>90%	80–90%	50–80%	30–50%	<30%
Cytotoxic scale	0 (zero)	1 (mild)	2 (moderate)	3 (intense)	4 (severe)

membrane, DNA synthesis, or metabolic activities as indicators of possible toxicity. Different approaches have been adopted such as neutral red assay to measure cell viability and damage in the cellular membrane, cellular proliferation by Coomassie blue, and the MTT or tetrazolium method for determination of mitochondrial function), and cellular injury by estimating lactate dehydrogenase.

The cytotoxicity of the BSs obtained from the *L. pentosus* assay was determined with the mouse fibroblast cell line [63]. The above-said cells were cultivated and supplemented with 10% of fetal bovine serum with 1% antibiotics. To determine the cytotoxicity of each BS was weighed and suspended at a concentration of 0.001-1 g/L. Subsequently, media containing fibroblasts cells at a concentration of 1×10^5 cells/mL was exposed to the test concentration. The cell value of cell viability and the highest level of cytotoxicity have been grouped as depicted in Table 6.5.

6.5   Phytotoxicity

One generally used bioassay for determination of toxicity is seed germination using phytotoxicity. Seed germination rate and elongation of the root can be considered as a rapid toxicity assessment approach, having various advantages, like sensitivity, inexpensive, rapid, and appropriateness for chemicals [66]. Phytotoxic effects can be determined on any seasonal seeds on a crop or plant. Approximately 20–30 seeds can be used after surface sterilization using sodium hypochlorite with subsequent washing with de-ionized water. Afterward, seeds can be grown in glass Petri plates containing 2–3 layers of laboratory filter paper soaked with the BSs tested. The above toxicity assessment can be done in a bacterial incubator at 37 °C. The test seeds can be periodically irrigated with a test solution containing BSs to maintain moisture. Consequently, the percentage of seed germination was determined as follows:

$$\%\textbf{seed germination} = \text{Number of seeds germinated}/\text{Total number of seeds} \times 100$$

On incubation, the length of the plant shoot and elongated root can be measured. The vigor index can also be determined as;

$$\textbf{Vigor index} = (\text{root length} + \text{shoot length}) \times \text{Seed germination percentage}$$

6.6 Conclusion and Future Directions

For the significant applications and commercialization of microbial surfactants and their inclusion in personal care and healthcare formulations, the explanation of safety and possible hazards datasheets for such BSs is necessary. In the past, immense research has been done about the production, optimization, and characterization of the BSs, but the data related to their toxicity fate, physicochemical behaviors, life cycle assessments, and degradable properties are inadequate. So, it is essential to increase the investigation about such concerns to elucidate such aspects, which are essential to get accurate safety data about their adoption to the industry. The major concerns and regulations about the surfactants are biodegradability, toxicity, and impact on human health. The adoption of microbial surfactants could expand all such aspects in comparison with synthetic surfactants and should be well-thought-out in the predictable future. As it was established by the EU regulation in 2005 that the use of completely or readily biodegradable surfactants should be adopted by the cosmetics and pharmaceuticals industry.

References

1. Banat IM, Franzetti A, Gandolfi I, Bestetti G, Martinotti MG, Fracchia L et al (2010) Microbial biosurfactants production, applications and future potential. Appl Microbiol Biotechnol 87 (2):427–444
2. Kumar P, Sharma PK, Sharma PK, Sharma D (2015) Micro-algal lipids: A potential source of biodiesel. JIPBS 2(2):135–143
3. Saharan BS, Sahu RK, Sharma D (2011) A review on biosurfactants: fermentation, current developments and perspectives. Genet Eng Biotechnol J 2011(1):1–14
4. Sharma D, Dhanjal DS, Mittal B (2017) Development of edible biofilm containing cinnamon to control food-borne pathogen. J Appl Pharm Sci 7(01):160–164
5. Waghmode S, Suryavanshi M, Sharma D, Satpute SK (2020) Planococcus species–an imminent resource to explore biosurfactant and bioactive metabolites for industrial applications. Front Bioeng Biotechnol 8:996
6. Rodrigues L, Banat IM, Teixeira J, Oliveira R (2006) Biosurfactants: potential applications in medicine. J Antimicrob Chemother 57(4):609–618
7. Karlapudi AP, Venkateswarulu TC, Tammineedi J, Kanumuri L, Ravuru BK, ramu Dirisala, V., & Kodali, V. P. (2018) Role of biosurfactants in bioremediation of oil pollution-a review. Petroleum 4(3):241–249
8. Sharma D, Saharan BS (eds) (2018) Microbial cell factories. CRC Press, Boca Raton

9. Rebello S, Anoopkumar AN, Sindhu R, Binod P, Pandey A, Aneesh EM (2020) Comparative life-cycle analysis of synthetic detergents and biosurfactants—an overview. In: Refining Biomass Residues for Sustainable Energy and Bioproducts. Academic Press, London, pp 511–521

10. Rebello S, Asok AK, Mundayoor S, Jisha MS (2014) Surfactants: toxicity, remediation and green surfactants. Environ Chem Lett 12(2):275–287

11. Bandala ER, Peláez MA, Salgado MJ, Torres L (2008) Degradation of sodium dodecyl sulphate in water using solar driven Fenton-like advanced oxidation processes. J Hazard Mater 151 (2–3):578–584

12. Azizullah A, Richter P, Jamil M, Häder DP (2012) Chronic toxicity of a laundry detergent to the freshwater flagellate Euglena gracilis. Ecotoxicology 21(7):1957–1964

13. Sharma V, Garg M, Devismita T, Thakur P, Henkel M, Kumar G (2018) Preservation of microbial spoilage of food by biosurfactantbased coating. Asian J Pharm Clin Res 11(2):98

14. Vecino X, Cruz JM, Moldes AB, Rodrigues LR (2017) Biosurfactants in cosmetic formulations: trends and challenges. Crit Rev Biotechnol 37(7):911–923

15. Das D, Dash U, Meher J, Misra PK (2013) Improving stability of concentrated coal–water slurry using mixture of a natural and synthetic surfactants. Fuel Process Technol 113:41–51

16. Singh J, Sharma D, Kumar G, Sharma NR (eds) (2018) MSicrobial bioprospecting for sustainable development. Springer, Cham

17. Tomislav I, Hrenovic J (2010) Surfactants in the environment. Arh Hig Rada Toksicol 61:95–110

18. Yuan CL, Xu ZZ, Fan MX, Liu HY, Xie YH, Zhu T (2014) Study on characteristics and harm of surfactants. J Chem Pharm Res 6(7):2233–2237

19. Franzetti A, Di Gennaro P, Bevilacqua A, Papacchini M, Bestetti G (2006) Environmental features of two commercial surfactants widely used in soil remediation. Chemosphere 62 (9):1474–1480

20. Sahnoun R, Mnif I, Fetoui H, Gdoura R, Chaabouni K, Makni-Ayadi F et al (2014) Evaluation of Bacillus subtilis SPB1 lipopeptide biosurfactant toxicity towards mice. Int J Pept Res Ther 20 (3):333–340

21. Moldes AB, Paradelo R, Rubinos D, Devesa-Rey R, Cruz JM, Barral MT (2011) Ex situ treatment of hydrocarbon-contaminated soil using biosurfactants from Lactobacillus pentosus. J Agric Food Chem 59(17):9443–9447

22. Manickam N, Bajaj A, Saini HS, Shanker R (2012) Surfactant mediated enhanced biodegradation of hexachlorocyclohexane (HCH) isomers by Sphingomonas sp. NM05. Biodegradation 23 (5):673–682

23. Bohnes FA, Laurent A (2019) LCA of aquaculture systems: methodological issues and potential improvements. Int J Life Cycle Assess 24(2):324–337

24. Thannimalay L, Yusoff S (2014) Comparative analysis of environmental evaluation of LAS and MES in detergent—A Malaysian case study. World Appl Sci J 31(9):1635–1647

25. Kopsahelis A, Kourmentza C, Zafiri C, Kornaros M (2018) Gate-to-gate life cycle assessment of biosurfactants and bioplasticizers production via biotechnological exploitation of fats and waste oils. J Chem Technol Biotechnol 93(10):2833–2841

26. Baccile N, Babonneau F, Banat IM, Ciesielska K, Cuvier AS, Devreese B et al (2017) Development of a cradle-to-grave approach for acetylated acidic sophorolipid biosurfactants. ACS Sustain Chem Eng 5(1):1186–1198

27. Hu X, Subramanian K, Wang H, Roelants SL, To MH, Soetaert W et al (2020) Guiding environmental sustainability of emerging bioconversion technology for waste-derived sophorolipid production by adopting a dynamic life cycle assessment (dLCA) approach. Environ Pollut 269:116101

28. Sharma D, Saharan BS, Kapil S (2016) Biosurfactants of lactic acid bacteria. Springer, Cham

29. Berna JL, Cassani G, Hager CD, Rehman N, López I, Schowanek D et al (2007) Anaerobic biodegradation of surfactants–scientific review. Tenside Surfactants Detergents 44(6):312–347

30. Garcia MT, Kaczerewska O, Ribosa I, Brycki B, Materna P, Drgas M (2016) Biodegradability and aquatic toxicity of quaternary ammonium-based gemini surfactants: effect of the spacer on their ecological properties. Chemosphere 154:155–160

31. Hagner M, Romantschuk M, Penttinen OP, Egfors A, Marchand C, Augustsson A (2018) Assessing toxicity of metal contaminated soil from glassworks sites with a battery of biotests. Sci Total Environ 613:30–38

32. Guhl W, Steber J (2006) The value of biodegradation screening test results for predicting the elimination of chemicals' organic carbon in waste water treatment plants. Chemosphere 63 (1):9–16

33. Li JL, Chen BH (2009) Surfactant-mediated biodegradation of polycyclic aromatic hydrocarbons. Materials 2(1):76–94

34. Li JL, Bai R (2005) Effect of a commercial alcohol ethoxylate surfactant (C 11-15 E 7) on. Biodegradation 16(1):57–65

35. Tiehm A (1994) Degradation of polycyclic aromatic hydrocarbons in the presence of synthetic surfactants. Appl Environ Microbiol 60(1):258–263

36. Garcia MT, Campos E, Sánchez-Leal J, Ribosa I (2006) Effect of linear alkylbenzene sulphonates (LAS) on the anaerobic digestion of sewage sludge. Water Res 40(15):2958–2964

37. Lima TM, Procópio LC, Brandão FD, Carvalho AM, Tótola MR, Borges AC (2011) Biodegradability of bacterial surfactants. Biodegradation 22(3):585–592

38. Hirata Y, Ryu M, Oda Y, Igarashi K, Nagatsuka A, Furuta T, Sugiura M (2009) Novel characteristics of sophorolipids, yeast glycolipid biosurfactants, as biodegradable low-foaming surfactants. J Biosci Bioeng 108(2):142–146

39. Renkin M (2003) Environmental profile of sophorolipid and rhamnolipid biosurfactants. Riv Ital Sostanze Grasse 80:249–252

40. Hogan DE, Tian F, Malm SW, Olivares C, Pacheco RP, Simonich MT et al (2019) Biodegradability and toxicity of monorhamnolipid biosurfactant diastereomers. J Hazard Mater 364:600–607

41. De Oliveira DW, Cara AB, Lechuga-Villena M, García-Román M, Melo VM, Gonçalves LR, Vaz DA (2017) Aquatic toxicity and biodegradability of a surfactant produced by Bacillus subtilis ICA56. J Environ Sci Health A 52(2):174–181

42. Santos DKF, Rufino RD, Luna JM, Santos VA, Sarubbo LA (2016a) Biosurfactants: multifunctional biomolecules of the 21st century. Int J Mol Sci 17(3):401

43. Santos DKF, Rufino RD, Luna JM, Santos VA, Sarubbo LA (2016b) Biosurfactants: multifunctional biomolecules of the 21st century. Int J Mol Sci 17(3):401

44. Anyanwu CU, Obi SKC, Okolo BN (2011) Lipopeptide biosurfactant production by Serratia marcescens NSK-1 strain isolated from petroleum-contaminated soil. Journal of Applied Sciences Research 7:79–87

45. Kiran GS, Priyadharsini S, Sajayan A, Priyadharsini GB, Poulose N, Selvin J (2017) Production of lipopeptide biosurfactant by a marine Nesterenkonia sp and its application in food industry. Front Microbiol 8:1138

46. Parthipan P, Preetham E, Machuca LL, Rahman PK, Murugan K, Rajasekar A (2017) Biosurfactant and degradative enzymes mediated crude oil degradation by bacterium Bacillus subtilis A1. Front Microbiol 8:193

47. Patowary K, Patowary R, Kalita MC, Deka S (2017) Characterization of biosurfactant produced during degradation of hydrocarbons using crude oil as sole source of carbon. Front Microbiol 8:279

48. Trejo-Castillo R, Martínez-Trujillo MA, García-Rivero M (2014) Effectiveness of crude biosurfactant mixture for enhanced biodegradation of hydrocarbon contaminated soil in slurry reactor. Int J Environ Res 8(3):727–732

49. Catter KM, Oliveira DFD, Sousa OVD, Gonçalves LRB, Vieira RHSDF, Alves CR (2016) Biosurfactant production by pseudomonas aeruginosa and burkholderia gladioli isolated from mangrove sediments using alternative substrates. Orbital: The Electronic Journal of Chemistry 8 (5):269–275

50. Fei D, Zhou GW, Yu ZQ, Gang HZ, Liu JF, Yang SZ et al (2020) Low-toxic and nonirritant biosurfactant Surfactin and its performances in detergent formulations. J Surfactant Deterg 23 (1):109–118

51. Zenati B, Chebbi A, Badis A, Eddouaouda K, Boutoumi H, El Hattab M et al (2018) A non-toxic microbial surfactant from Marinobacter hydrocarbonoclasticus SdK644 for crude oil solubilization enhancement. Ecotoxicol Environ Saf 154:100–107

52. Sharma, D., Saharan, B. S., Chauhan, N., Bansal, A., & Procha, S. (2014). Production and structural characterization of Lactobacillus helveticus derived biosurfactant Sci World J, 2014

53. Saravanakumari P, Mani K (2010) Structural characterization of a novel xylolipid biosurfactant from Lactococcus lactis and analysis of antibacterial activity against multi-drug resistant pathogens. Bioresour Technol 101(22):8851–8854

54. Rodríguez-López L, Rincón-Fontán M, Vecino X, Cruz JM, Moldes AB (2019) Preservative and irritant capacity of biosurfactants from different sources: A comparative study. J Pharm Sci 108(7):2296–2304

55. Arwa BR, Vladimir BB (2016) In silico toxicology: computational methods for the prediction of chemical toxicity. Wiley Interdiscip Rev Comput Mol Sci 6:147–172

56. Clemedson C, Barile FA, Chesne C, Cottin M, Curren R, Eckwall B, Ferro M, Gomez-Lechon MJ, Imai K, Janus J, Kemp RB, Kerszman G, Kjellstrand P, Lavrijsen K, Logemann P, McFarlane-Abdulla E, Roguet R, Segner H, Thuvander A, Walum E, Ekwall B (2000) MEIC evaluation of acute systemic toxicity. Part VII. Prediction of human toxicity by results from testing of the first 30 reference chemicals with 27 further in vitro assays. ATLA 28:159–200

57. Maheshwari DG, Shaikh NK (2016) An overview on toxicity testing method. Int J Pharm Technol 8(2):3834–3849

58. Hayes AW, Loomis TA (1996) Loomis's essentials of toxicology. Elsevier, Amsterdam

59. Diniz Rufino R, Moura de Luna J, de Campos Takaki GM, Asfora Sarubbo L (2014) Characterization and properties of the biosurfactant produced by Candida lipolytica UCP 0988. Electron J Biotechnol 17(1):6–6

60. Bezerra de Souza Sobrinho H, de Luna JM, Rufino RD, Figueiredo Porto AL, Sarubbo LA (2013) Assessment of toxicity of a biosurfactant from Candida sphaerica UCP 0995 cultivated with industrial residues in a bioreactor. Electron J Biotechnol 16(4):4–4

61. Deivakumari M, Sanjivkumar M, Suganya AM, Prabakaran JR, Palavesam A, Immanuel G (2020) Studies on reclamation of crude oil polluted soil by biosurfactant producing Pseudomonas aeruginosa (DKB1). Biocatal Agric Biotechnol 29:101773

62. Apha A (2005) Standard methods for the examination of water and wastewater, vol 21. WEF, 2005, Washington, D.C., pp 258–259

63. Rodríguez-López L, López-Prieto A, Lopez-Álvarez M, Pérez-Davila S, Serra J, González P et al (2020) Characterization and cytotoxic effect of biosurfactants obtained from different sources. ACS omega, Washington

64. López-Prieto A, Vecino X, Rodríguez-López L, Moldes AB, Cruz JM (2019) A multifunctional biosurfactant extract obtained from corn steep water as bactericide for agrifood industry. Foods 8(9):410

65. Vecino X, Barbosa-Pereira L, Devesa-Rey R, Cruz JM, Moldes AB (2015) Optimization of liquid–liquid extraction of biosurfactants from corn steep liquor. Bioprocess Biosyst Eng 38 (9):1629–1637

66. Gaur VK, Regar RK, Dhiman N, Gautam K, Srivastava JK, Patnaik S et al (2019) Biosynthesis and characterization of sophorolipid biosurfactant by Candida spp.: Application as food emulsifier and antibacterial agent. Bioresour Technol 285:121314